DIGITAL IMAGE PROCESSING

An Algorithmic Approach
with MATLAB®

CHAPMAN & HALL/CRC
TEXTBOOKS IN COMPUTING

DIGITAL IMAGE PROCESSING

An Algorithmic Approach with MATLAB®

Uvais Qidwai *and* C. H. Chen

CRC Press
Taylor & Francis Group
Boca Raton   London   New York

CRC Press is an imprint of the
Taylor & Francis Group , an **informa** business

A CHAPMAN & HALL BOOK

MATLAB® and Simulink® are trademarks of The MathWorks, Inc. and are used with permission. The Math-Works does not warrant the accuracy of the text or exercises in this book. This book's use or discussion of MATLAB® and Simulink® software or related products does not constitute endorsement or sponsorship by The MathWorks of a particular pedagogical approach or particular use of the MATLAB® and Simulink® software.

Chapman & Hall/CRC
Taylor & Francis Group
6000 Broken Sound Parkway NW, Suite 300
Boca Raton, FL 33487-2742

First issued in paperback 2017

© 2010 by Taylor and Francis Group, LLC
Chapman & Hall/CRC is an imprint of Taylor & Francis Group, an Informa business

No claim to original U.S. Government works

ISBN 13: 978-1-138-11518-7 (pbk)
ISBN 13: 978-1-4200-7950-0 (hbk)

Library of Congress Cataloging-in-Publication Data

Qidwai, Uvais.
    Digital image processing : an algorithmic approach with MATLAB / authors, Uvais Qidwai, C.H. Chen.
        p. cm. -- (Textbooks in computing)
        "A CRC title."
        Includes bibliographical references and index.
        ISBN 978-1-4200-7950-0 (hardcover : alk. paper)
        1. Image processing--Digital techniques. 2. MATLAB. I. Chen, C. H. (Chi-hau), 1937- II. Title. III. Series.

TA1637Q54 2010
621.36'70285--dc22   2009032785

Visit the Taylor & Francis Web site at
http://www.taylorandfrancis.com

and the CRC Press Web site at
http://www.crcpress.com

Table of Contents

Preface

Why another book on image processing? One might wonder, especially when almost all of the books available in the market are written by very well-versed and experienced academicians. Even more intriguing is the fact that I am a lot younger compared to all of them when they wrote those books! However, I think this is the main driving force behind this effort. Over the past few years when I have been involved teaching the subject in various countries around the world, I have felt that the available textbooks are not very "student friendly." Not too long ago, I shared similar feelings when I was on the student benches myself. In today's ultra-fast-paced life, the definition of "student friendly" is predominantly related to how fast the information can be disseminated to the students in as easy (and fun) way as possible. This definition, essentially, depicts the whole intent of writing this book.

This book covers topics that I believe are essential for undergraduate students in the areas of engineering and sciences in order to obtain a minimum understanding of the subject of digital image processing. At the same time, the book is written keeping in mind "average" students not only in the United States but elsewhere in the world as well. This is also the reason that the book has been proposed as a textbook for the subject since I believe that a textbook must be completely (or at least 90%) comprehensible by the students. However, students who want to delve deeper into the topics of the book can refer to some of the references in the bibliography section including several Web links. The book can also be a very good starting point for student projects as well as for start-up research in the field of image processing because it will give an encouraging jump-start to students without bogging them down with syntactical and debugging issues that they might encounter when using a programming environment other than MATLAB®, or even trying out MATLAB for the first time for imaging applications.

The magic number of 15 chapters is based on a typical 15-week semester (plus or minus two more for the exams, etc.). Hence, typically one chapter can be completed per week, although in some cases, it may spill over to the next week as well. Each chapter is divided into three distinct sections. Their content varies in length relative to the topic being covered. The first of these sections is related to the actual theoretical contents to be covered under the chapter title. These theoretical topics are also presented in a very simple and basic style with generic language and mathematics. In several places, only a final result has been presented rather than the complex mathematical derivation of that result. The intent of this section is to equip the student with general understanding of the topic and any mathematical tool they will be using.

The second section (explicitly titled "Algorithmic Account") explains the theoretical concepts from the theoretical section in the form of a flowchart to streamline the concepts and to lay a foundation for students to get ready for coding in any programming language. The language used in the flowchart is purposely kept simple and generic, and standard symbols for flowcharts are used. The third section ("MATLAB Code") will complete this understanding by providing the actual MATLAB code for realizing the concepts and their applications. Specific emphasis is given on reproducing the figures presented in the chapter through the listed code in this section. At the end of each chapter, a bulleted summary of the chapter is provided. This gives a bird's-eye view to the students as well as the instructors of the topics covered in the chapter. The exercises at the end of the chapter are mostly programming based so that students learn the underlying concepts through practice.

By no means can I claim that this is sufficient for students to become well-versed in the area of image processing. It will, however, open the door to a fundamental understanding, and make it very easy for them afterward to comprehend the advanced topics in the field, as well as other mathematical details.

The book has some additional support material that can be found on the following Web site:

http://faculty.qu.edu.qa/qidwai/DIP

It contains the following items:

- PowerPoint slides that can be used for chapterwise lectures

- A GUI tool infrastructure in MATLAB that can be developed by the student into a full-functionality image processing GUI tool as a course project

- A folder containing all the images used in the book with MATLAB codes

In order to gain full benefit from the book, one must have MATLAB 6.5 or higher with toolboxes on image processing, image acquisition, statistics, signal processing, and fuzzy logic.

Disclaimer: To the best of my knowledge, all of the text, tables, images, and codes in the book are either original or are taken from public domain Web sites from the Internet. The images football.jpg, circles.png, coins.png, and testpat1.png are reproduced with permission from The MathWorks Inc. (U.S.), and onion.png and peppers.png are reproduced with permission of Jeff Mather, also of MathWorks.

MATLAB° is a registered trademark of The MathWorks, Inc. For product information, please contact:

The MathWorks, Inc.
3 Apple Hill Drive
Natick, MA 01760-2098 USA
Tel: 508-647-7000
Fax: 508-647-7001
E-mail: info@mathworks.com
Web: www.mathworks.com

Uvais Qidwai
April 2009

About the Authors

Uvais Qidwai received his Ph.D. from the University of Massachusetts–Dartmouth in 2001, where he studied in the electrical and computer engineering department. He taught in the electrical engineering and computer science department at Tulane University, in New Orleans, as an assistant professor, and was in charge of the robotics lab from June 2001 to June 2005. He joined the computer science and engineering department, Qatar University, in the Fall of 2005 as an assistant professor. His present interests in research include robotics, image enhancement and understanding for machine vision applications, fuzzy computations, signal processing and interfacing, expert system for testing pipelines, and intelligent algorithms for medical informatics. He has participated in several government- and industry-funded projects in the United States, Saudi Arabia, Qatar, and Pakistan, and has published over 55 papers in reputable journals and conference proceedings. His most recent research is related to target tracking in real-time video streams for sports medicine, robotic vision applications for autonomous service robots, and development of new data filters using imaging techniques.

C. H. Chen received his Ph.D. in electrical engineering from Purdue University in 1965. He has been a faculty member with the University of Massachusetts–Dartmouth since 1968 where he is now chancellor professor. He has taught the digital image processing course since 1975. Dr. Chen was the associate editor of *IEEE Transactions on Acoustics, Speech, and Signal Processing* from 1982 to 1986, and associate editor on information processing for remote sensing of *IEEE Transactions on Geoscience and Remote Sensing* 1985 to 2000. He is an IEEE fellow (1988), life fellow (2003), and also a fellow of the International Association of Pattern Recognition (IAPR 1996). Currently, he is an associate editor of the *International*

Journal of Pattern Recognition and Artificial Intelligence (since 1985), and on the editorial board of *Pattern Recognition* (since 2009). In addition to the remote sensing and geophysical applications of statistical pattern recognition, he has been active with the signal and image processing of medical ultrasound images as well as industrial ultrasonic data for non-destructive evaluation of materials He has authored 25 books in his areas of research interest. Two of his edited books recently published by CRC Press are *Signal Processing for Remote Sensing,* 2007 and *Image Processing for Remote Sensing,* 2007.

CHAPTER 1

Introduction to Image Processing and the MATLAB® Environment

1.1 INTRODUCTION

T HIS CHAPTER BRIEFLY INTRODUCES the scope of image processing. Modern digital technology has made it possible to manipulate multi-dimensional signals with systems that range from simple digital circuits to advanced parallel computers. The goal of this manipulation can be divided into three main categories and several subcategories:

- Image processing
  - Image input and output
  - Image adjustments (brightness, contrast, colors, etc.)
  - Image enhancement
  - Image filtering
  - Image transformations
  - Image compression
  - Watermarking and encryption

- Image analysis
  - Image statistics
  - Binary operations
  - Region of interest operations
- Image understanding
  - Image classification
  - Image registration
  - Image clustering
  - Target identification and tracking

We will focus on the fundamental concepts of the main categories to the extent needed by most engineering curricula requirements. Occasionally, advanced topics as well as open areas of research will be pointed out. Further, we will restrict ourselves to two-dimensional (2D) image processing, although most of the concepts and techniques that are to be described can be extended easily to three or more dimensions.

1.1.1 What Is an Image?

A digital image is a 2D signal in essence, and is the digital version of the 2D manifestation of the real-world 3D scene. Although the words *picture* and *image* are quite synonymous, we will make the useful distinction that "picture" is the analog version of "image." An image is a function of two real variables, for example, $a(x,y)$ with a as the amplitude (e.g., brightness) of the image at the *real* coordinate position (x,y). An image may be considered to contain subimages, which are sometimes referred to as *regions* or *regions of interest* (ROIs). The amplitudes of a given image will almost always be either real numbers or integers. The latter is usually the result of a quantization process that converts a continuous range (say, between 0 and 100%) to a discrete number of levels.

1.2 DIGITAL IMAGE DEFINITIONS: THEORETICAL ACCOUNT

A digital image $a[m,n]$ described in a 2D discrete space is derived from an analog image $a(x,y)$ in a 2D continuous space through a sampling process that is frequently referred to as *digitization*. Each sample of the image is called a *pixel* (derived somewhere down the line from *picture element*). The manner in which sampling has been performed can affect the size and details present in the image. Some of these effects of digitization are shown in Figure 1.1.

The 2D continuous image $a(x,y)$ is divided into N rows and M columns. The intersection of a row and a column is termed a *pixel*. The value assigned to the integer coordinates $[m,n]$ with $\{m = 0, 1, 2, ..., M–1\}$ and $\{n = 0, 1, 2, ..., N–1\}$ is $a[m,n]$. In fact, in most cases $a(x,y)$, which we might consider to be the physical signal that impinges on the face of a 2D sensor, is actually a function of many variables, including depth (z), color (λ), and time (t).

Images can be of various sizes; however, there are standard values for the various parameters encountered in digital image processing. These values occur due to the hardware constraints caused by the imaging source and/or by certain standards of imaging protocols being used. For instance, some typical dimensions of images are 256×256, 640×480, etc. Similarly, the grayscale values of each pixel, G, are also subjected to the constraints imposed by quantizing hardware that converts the analog picture value into its digital equivalent. Again, there can be several possibilities for the range of these values, but frequently we see that it depends on the number of bits being used to represent each value. For several algorithmic reasons, the number of bits is constrained to be a power of 2, that is, $G = 2^B$, where B is the number of bits in the binary representation of the brightness levels. When $B > 1$, we speak of a *gray-level image*; when $B = 1$, we speak of a *binary image*. In a binary image, there are just two gray levels, which can be referred to, for example, as "black" and "white" or "0" and "1." This notion is further facilitated by digital circuits that handle these values or by the use of certain algorithms such as the (fast) Fourier transform.

In Figure 1.1, images in (b), (c), and (e) have the same size as the original. The only difference is that the number of pixels being skipped (and replaced with zeros) is different: 10-, 2-, and 5- pixel spacing, respectively. Obviously, the larger the spacing, the more information is lost, and combining the actual sampled points will result in a smaller image, as shown in the images in (d) and (f). One more observation on pixel loss can be made through the images in (d) and (f), where the original image is much more distorted for larger sampling spacing. The effects shown in Figure 1.1 can also be defined in technical terms: spatial resolution, which describes the level of detail an image holds within its physical dimensions. Essentially, this means the number of detail elements (or pixels) present in the rows

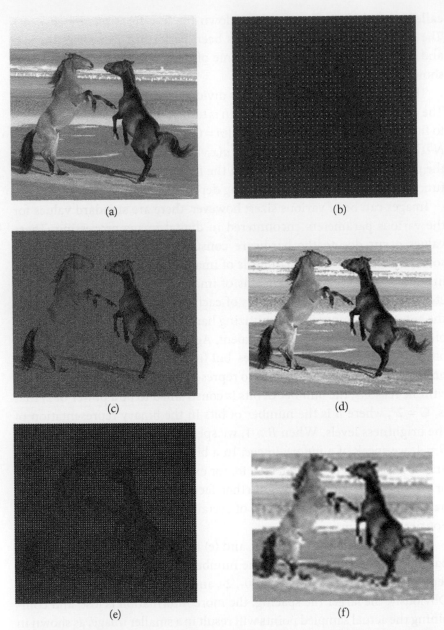

(a) (b)

(c) (d)

(e) (f)

FIGURE 1.1 (See color insert following Page 204.) Digitization of a continuous image. (a) Original image of size 391 × 400 × 3, (b) image information is almost lost if sampled at a distance of 10 pixels, (c) resultant pixels when sampling distance is 2, (d) new image with 2-pixel sampling with size 196 × 200 × 3, (e) resultant pixels when sampling distance is 5, (f) new image with 5-pixel sampling with size 79 × 80 × 3.

and columns of the image. Higher resolution means more image detail. Consequently, image resolution can be tied to physical size (e.g., lines per millimeter, lines per inch) or to the overall size of a picture (lines per picture height, number of pixels, and pixel density).

1.3 IMAGE PROPERTIES

1.3.1 Signal-to-Noise Ratio

Signal-to-noise ratio (SNR) is an important parameter to judge, analyze, and classify several image processing techniques. As described previously, in modern camera systems the noise is frequently limited by the following:

- Amplifier noise in the case of color cameras

- Thermal noise, which itself is limited by the chip temperature K and the exposure time T

- Photon noise, which is limited by the photon production rate and the exposure time T

Effectively, SNR is calculated as

$$SNR = \frac{Signal\_Power}{Noise\_Power} = 20\log\frac{P_{Signal}}{P_{Noise}}. \tag{1.1}$$

How noise can affect the image is shown in Figure 1.2. The added noise is of "salt-and-pepper" type. The addition of noise is only the addition of randomness to the clean pixel values. This, however, is a very common problem in digital image acquisition, where the randomness appears from hardware elements. Chip behavior can be random on account of thermal conditions; such behavior is inherently a charge flow phenomenon and is highly dependent on the temperature.

1.3.2 Image Bit Resolution

Image bit resolution, or simply *image resolution,* refers to the number of grayscale levels or the number of pixels present in the image. Compared to the term *spatial resolution* mentioned earlier (following Figure 1.1), the commonly used terminology of *image resolution* refers to the representational capability of the image, using a certain number of bits to represent

(a)

(b)

FIGURE 1.2 The effect of noise on images. (a) Original Image, (b) noisy image with signal-to-noise-ratio (SNR) 20 dB.

the intensity levels at various points in the image. A more quantitative and algorithmic description is given in Section 1.5.3 and Figure 1.4(c) of this chapter. The effect of reducing image resolution (either number of gray levels or number of pixels) can be explored to understand these important variables. Note that the ability to recognize small features and locate boundaries requires enough pixels, and that too few gray levels produce visible "contouring" artifacts in smoothly varying regions. These are shown in Figure 1.3.

FIGURE 1.3 (See color insert following Page 204.) The effect of bit resolution. (a) Original image with 8-bit resolution, (b) 4-bit resolution, (c) 3-bit resolution, (d) 2-bit resolution.

1.4 MATLAB

MATLAB stands for MATrix LABoratory, and it is a software environment extremely suitable for engineering algorithmic development and simulation applications. Commercialized in 1984 by The MathWorks Inc. (Natick, MA), the MATLAB project was initiated as a result of the recognition of the fact that engineers and scientists need a more powerful and productive computation environment beyond those provided by languages such as C and FORTRAN. Since then, MATLAB has been heavily extended and has become a de facto standard for scientific and engineering calculations, visualization, and simulations. Essentially, it is a data analysis and visualization tool that has been designed with powerful support for matrices and matrix operations. It operates as an interactive programming environment. MATLAB is well adapted to numerical experiments because the underlying algorithms for MATLAB's built-in functions and supplied m-files are based on the standard libraries LINPACK and EISPACK. MATLAB programs and script files always have file names ending with

".m"; the programming language is exceptionally straightforward because almost every data object is assumed to be an array. Graphical output is available to supplement numerical results.

Although MATLAB is claimed to be the optimal tool for working with matrices, its performance can deteriorate significantly if it is not used carefully. For instance, several algorithmic tasks involve the use of loops to perform different types of iterative procedures. It turns out that MATLAB is not too loop friendly. However, if the number of loops can be reduced and most of the operations can be converted into matrix-based manipulations, then the performance suddenly improves greatly. For instance, calculation of the sum-square of a row vector A can be much more efficient in MATLAB if performed as $A*A'$ instead of having to go through a loop within which a running sum is calculated.

MATLAB has a basic computation engine that is capable of doing all the wonderful things in terms of computations. It does that by utilizing basic computational functions. Some of these functions are open so that their code can be read. However, the source code for most of the main built-in functions, such as the fft() function, cannot be read. These built-in functions are the keys to the powerful computational structure which MATLAB has and the people who have developed it consider this as one of their main assets. There are a number of other helping files that surround this core of MATLAB, and these files include the help documentation, compilers for converting MATLAB files into C or JAVA, and several other operating-system-dependent libraries. MathWorks also enhanced the power of MATLAB by introducing sister software or engines such as SIMULINK® and Real-Time Workshop. However, all of this would have made MATLAB a wonderful mathematical tool only. The real power of this environment comes from a huge number of specialized functions called *toolboxes*, specifically written by experts in a certain area not necessarily related to programming. These toolboxes include specialized functions and related files to perform the high-level operations related to that particular field. There were approximately 80 toolboxes, each targeting a specific area of interest at the time these lines were written! And these are only those toolboxes that are recognized by MathWorks. Around the world, many graduate students develop a toolbox of their own related to their thesis or research work. Some of the interesting toolboxes of use to us within the context of this book are Images, Signal, Comm, Control, Imaq, Fuzzy, Ident, Nnet, Stats, and Wavelet.

1.4.1 Why MATLAB for Image Processing

As explained earlier, an image is just a set of values organized in the form of a matrix. Because MATLAB is optimal for matrix operations, it makes a lot of sense to use MATLAB for image-related operations. For most of the processing needs, we will be using MATLAB's Image Processing Toolbox (images). Other than real-time acquisition of images, this toolbox has all the necessary tools to perform various geometric, arithmetic, logical, as well as other higher level transformations on the images. The toolbox is also capable of handling both color and grayscale images. However, most of the image processing is focused on grayscale images. Even for the color images, the processing is done on the converted image, which is obtained by mapping the RGB color space into grayscale space. These concepts will be discussed in detail in Chapter 2.

Although structural compatibility is a great motivation to use MATLAB for image processing, there are several other reasons for doing so. Most researchers in the area of image processing use MATLAB as their main platform for software implementation, which thus gives a common language to compare different algorithms and designs. Also, the speeds of various algorithms may also be compared on a common platform, which would be difficult if different people were using different programming languages that vary considerably in terms of speed of operations. Another interesting reason to use MATLAB, most interesting to those engineering students who do not like a lot of coding, is the brevity of code in MATLAB. Table 1.1 compares some examples of basic operations in MATLAB and C as a reference. One can imagine from the comparison how dramatic the difference will be for complex operations such as convolution, filtering, and matrix inversion. Convolution is the heart of almost all of the filtering and time and frequency domain transformations in image processing, and must be done as fast as possible. The MATLAB function conv() is also one of the well-kept secrets of MathWorks and is the heart of these operations. The function has been optimized for the matrix operations and, hence, operations in MATLAB become faster and more efficient than coding in other languages.

Of course, there is a limit to the usefulness of MATLAB. Although it is wonderful for algorithmic testing, it is not very suitable for real-time imaging applications due to the slowness of processing. This slowness comes from more levels of compilation and interpretation compared to the other languages, as well as from iterative procedures where loops are used frequently in real-world

TABLE 1.1   Comparison of MATLAB and C Code for Simple Matrix Operations

Operation	Part of C code	MATLAB statements
Addition of two matrices A and B	```for (i==1, i<=M, i++)
{
for (j==1, j<=N, j++)
{
D[i][j]=A[i][j]+B[i][j];
{
}```	D = A + B;
Multiplication of two matrices A and B	```for (i==1, i<=M, i++)
{
for (j==1, j<=N, j++)
{
for (c==1, c<=N, c++)
{
for (r==1, r<=M, r++)
{
D[i][j]+=A[i][c]*B[j][r];
}
}
}
}```	D = A * B;

applications. The best solution would be to test the logic and algorithms in MATLAB first because this will boost the initial development speed. Once the algorithm is finalized, it should be translated into C and then compiled into an executable file for real-time or near real-time applications. One comment for those students who are not too enthusiastic about coding: "You will forget about other languages once you start working with MATLAB." This is true most of the time for typical engineering student chores, as observed from personal experience. However, one must realize the fact that for real-world applications, the need for a near machine-language-type environment cannot be eliminated, at least for the time being.

1.4.2 The Image Processing Toolbox in MATLAB

The Image Processing Toolbox in MATLAB is a collection of functions that supports a wide range of image processing operations, including the following:

- Image I/O
- Spatial transformations

- Morphological operations

- Neighborhood and block operations

- Filter design and image enhancement

- Image registration

Many of the toolbox functions are MATLAB m-files, a series of MATLAB statements that implement specialized image processing algorithms. One can extend the capabilities of the Image Processing Toolbox by writing customized m-files, or by using the toolbox in combination with other toolboxes, such as the Signal Processing Toolbox and the Wavelet Toolbox, etc. The toolbox is also supported by a full complement of demo applications. These are very useful as templates for learning and customizing the applications. To view all the Image Processing Toolbox demos, call the iptdemos function. This displays an HTML page in the MATLAB Help browser that lists all the Image Processing Toolbox demos.

The toolbox demos are located under the subdirectory MATLAB\ toolbox\images\imdemos.

1.5 ALGORITHMIC ACCOUNT

In this section, the algorithmic logic behind the operations shown in Figures 1.1–1.3 is presented in the form of flow charts. Once understood, the same results can be obtained by coding in any programming language of choice.

1.5.1 Sampling

The effect of sampling, as shown in Figure 1.1, is mainly the reduction of number of pixels present in the image, which results in some loss of information in the image and manifests itself as quantization and apparent boxlike patterns in the image. The images in Figure 1.1 were created using the logic shown in Figure 1.4(a).

1.5.2 Noisy Image

Figure 1.2(b) represents the noisy image, which was obtained by adding the clean image with a specific type of noise called *salt and pepper*, and will be discussed in detail in Chapter 4. The methodology used in generating the

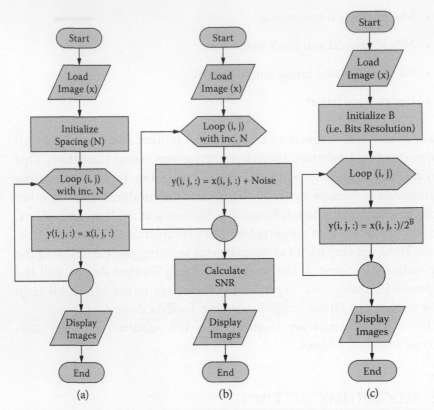

FIGURE 1.4 Programming logic for generating the images shown in Figures 1.1–1.3. (a) Sampling, (b) noisy image and SNR, (c) bit resolution.

images of Figure 1.2 is shown in Figure 1.4(b). The word *noise* is general in the diagram and can be defined as per requirements. The noise is a collection of random numbers, and each of these is added to the clean image. The signal-to-noise ratio (SNR) is calculated as in Equation 1.2.

$$snr = 20\log\frac{\sum_i\sum_j\sum_k x_{ijk}^2}{\sum_i\sum_j\sum_k (y-x)_{ijk}^2}. \tag{1.2}$$

1.5.3 Bit Resolution

Bit resolution refers to the number of bits needed to represent the distinct grayscale levels present in the images. Since the maximum value on any layer of the image is 255 in an 8-bit representation system (which is the default system in Windows), dividing each value by a certain power of 2 will result in reducing the grayscale value and, consequently, the number of bits needed to represent this value will be smaller. After the division, a rounding operation produces further quantization by removing the fractional part from the pixel values. Figure 1.4(c) depicts this strategy.

1.6 MATLAB CODE

In this section, first a brief "jump-start" is provided to the user to get going with the Image Processing Toolbox and, second, code is provided for the image operations presented in Figures 1.1–1.3.

1.6.1 Basic Steps

Clear the MATLAB workspace of any variables and close open figure windows:

```
Clear all; close all;
```

Although not essential, it is always a good practice to include this step in the beginning of an m-file that is being executed from start and is not dependent on the previous data. To read an image, use the imread command:

```
I = imread('horses.jpg');
```

infers from the file that the graphics file format is JPEG and the name is "horses," and stores it into the variable I. For the list of supported graphics file formats, see the imread function reference documentation.

The image can then be displayed by using one of the two image display functions: imshow or imview. You can use either one to display an image:

```
imshow(I)
```

Your choice of which display function to use depends on what you want to do. For example, because it displays the image in a MATLAB figure window, imshow provides access to figure annotation and printing capabilities. The imview function displays an image in a separate Java-based window called the Image Viewer, which provides access to additional tools that aid in navigating around an image, especially large images, and enable the inspection of pixels in an image. Figure 1.1(a) is the result of execution of the aforementioned command `imshow()`. The MATLAB code for the images in Figures 1.1–1.3 is

1.6.2 Sampling

```
clear all ; close all
x = imread('horses.jpg');
imshow(x)
[r,c,s] = size(x) ;
N = 10 ;
y = zeros(size(x)) ;
y(1:N:r,1:N:c,:) = x(1:N:r,1:N:c,:) ;
figure
imshow(y*0.01)
z = y(1:N:r,1:N:c,:) ;
figure
imshow(z*0.005)
```

1.6.3 Noisy Image

```
clear all ; close all
x = imread('flowers001.jpg');
imshow(x)
[r,c,s] = size(x) ;
y = imnoise(x,'salt & pepper',0.19);
snr = 20*log10(sum(sum(sum(x.^2)))/
sum(sum(sum((y-x).^2)))) ;
figure
imshow(y)
```

1.6.4 Bit Resolution

```
clear all ; close all
x = imread('valley.jpg');
imshow(x)
[r,c,s] = size(x) ;
m = max(max(max(x))) ;
b = [7 6 5] ;
```

```
for i = 1 : length(b)
    d = 2^b(i) ;
    z = round(x/d) ;
    figure
    imshow(z*d)
end
```

1.7 SUMMARY

- Images are defined as two-dimensional digital signals.

- Images are formed by capturing the color intensity values from the real-world 3D views.

- Each image element is called a *pixel*.

- Each pixel is a function of its position in the image. For instance, $x(i,j)$ represents a value of the light intensity at position ith row and jth column of the image.

- The values of these pixels suffer degradation from the procedures of image representation as well as from external factors.

- Image processing is the general name given to a number of operations, including image acquisition, image representation, image enhancement, image compression, image classification, etc.

- Once acquired, the image is represented as a matrix in computer memory and can be of 2 or higher dimensions.

- Color images are usually treated as compound images with three layers corresponding to the three electron beams in the display hardware of the monitor screen.

- MATLAB is a professional-grade computational environment developed by MathWorks Inc, Natick, MA. Because it is optimized in terms of operations and manipulations for matrices, the images become the part of its data space quite naturally.

- MATLAB can reduce several computationally complicated procedures needed for image processing to simpler code. This decreases the development time, and more emphasis can be given to algorithmic development rather than to coding.

- MATLAB has a number of collections of dedicated functions called toolboxes, which are specific to a particular area of computation and are usually developed by the experts in that field.

- Included code and its logic have been used to demonstrate some simple image manipulation strategies.

- The effects of sampling, number of bits needed to represent a grayscale value in an image, cause modifications to the pixel values of the resulting image.

- Sampling an image at a pixel distance greater than 1 results in a smaller image with some loss of information. The loss increases with increasing sampling distance, but the image size also decreases accordingly.

- Bit resolution of an image will affect the number of distinct grayscale levels present in an image.

- Typically, in the Windows environment, 8 bits are used to represent grayscale values, and three layers of such bit sets are used to represent colors.

- Noise in an image corresponds to the randomness present in pixel values. In real life, this can happen due to a number of factors such as randomly increased electron emissions due to thermal variations.

1.8 EXERCISES

1. Use the code for the noisy image to implement three different types of noise while keeping the SNR constant. Then, by visual inspection only, grade the quality of the images as bad, worse, and worst. The three types of noise you may try out are the Gaussian, uniform, and Poisson distributions. The fixed SNR is 10 dB.

2. Can an image with 4-bit representation be improved in visual quality by representing the same grayscale levels with 8 bits? Prove your answer with MATLAB code. What convinced you that the quality improved or degraded?

3. One of the methods for improving the pixelization (or the box formation) effect is to increase the size of the image by inserting the average values of the neighboring pixels between these pixels. Use the

image from Figure 1.1(f), increase its size by a factor of 3; comment on the quality of this new image. You must use the appropriate code provided in the chapter, and modify it to meet the requirements of this question.

4. If a noisy image is subsampled (i.e., the sampling distance is greater than 1), will this affect the noise content of the image visually? Will this affect the SNR? Prove your answers through MATLAB code.

5. Repeat the code for sampling with the modification that the new image is not initialized to zero in the beginning. You must initialize it to 255 (all white pixels) and compare the quality of the resulting sampled images for sampling distances of 2, 5, and 10 with those in Figure 1.1.

CHAPTER 2

Image Acquisition, Types, and File I/O

2.1 IMAGE ACQUISITION

IMAGE PROCESSING BEGINS WITH the image; hence, image acquisition is the first key step to investigate. Both the hardware and software used for acquisition can appear quite different, depending on the type of image source and interface. This can vary from video cameras with appropriate frame grabbers (either on a card internal to the computer or an external device connected to it by a parallel or serial interface), to digital cameras with their own serial or SCSI interfaces, to slide or flatbed scanners usually with an SCSI or USB interface. Connecting this hardware properly and installing the appropriate software drivers is sometimes confusing, particularly under Windows. The following is an abridged list of common sources for image acquisition:

- Real-world 3D objects are converted into digital images
  - Cameras (CCD, CID, CMOS)
  - Infrared cameras
  - Ultrasounds
  - X-rays and radiography
  - Magnetic resonance imaging (MRI)
  - Satellite imagers
- Scanners

2.1.1 Cameras

The cameras and recording media available for modern digital image processing applications are changing at a significant pace. Various types of technologies are available in the market, such as charge-coupled device (CCD), charge injection device (CID), and complementary metal oxide semiconductor (CMOS) cameras. The techniques that are used to characterize the basic type of CCD camera remain "universal" and can be transparently applied to other technologies. The basic principle of any charge-based camera is the emission of charges proportional to the incident light on these cells from the photosensitive elements or cells present in the main imaging hardware.

Whereas regular cameras generate images based on the photoemissions of sensors in the camera, infrared cameras exploit similar behavior of thermal cells; hence, they can detect "heat waves" within certain temperature ranges and convert them into electrical signals, which are then displayed as images. If the same principle is modified with radiation sensors, then a camera would detect the radioactive emissions and generate the equivalent electrical signals. This constitutes the basis of generation of digital x-rays and other radiographic images. All of these cameras are still direct imaging methods, in which the image is obtained directly from the source object without any mapping or transformation of the data space. Figure 2.1 shows some of the commonly used cameras for various applications.

However, ultrasonic and MRI images are not obtained so easily. They have to go through massive mapping and transformation operations before they can be seen. A typical ultrasonic image is based on the reflections of sound waves from the nonhomogeneities in its path. These reflections are recoded with precise time-of-flight (ToF) information related to the arrival of this reflection at the sensor. One such waveform in 1D is called *A-scan*. If the ultrasonic transceiver is now shifted and a new reflection is obtained, then this will be another A-scan. This procedure is repeated for the whole area to be imaged, and a huge data cube is obtained. Taking slices out of this cube perpendicular to the direction of the input ultrasonic pulse, and assigning normalized values to the various points present in the slice will appear as the cross-sectional image of the data cube and is called C-scan. Figure 2.2 shows the typical ultrasonic scanning principle and the resulting image as used in ultrasonic nondestructive testing of metals.

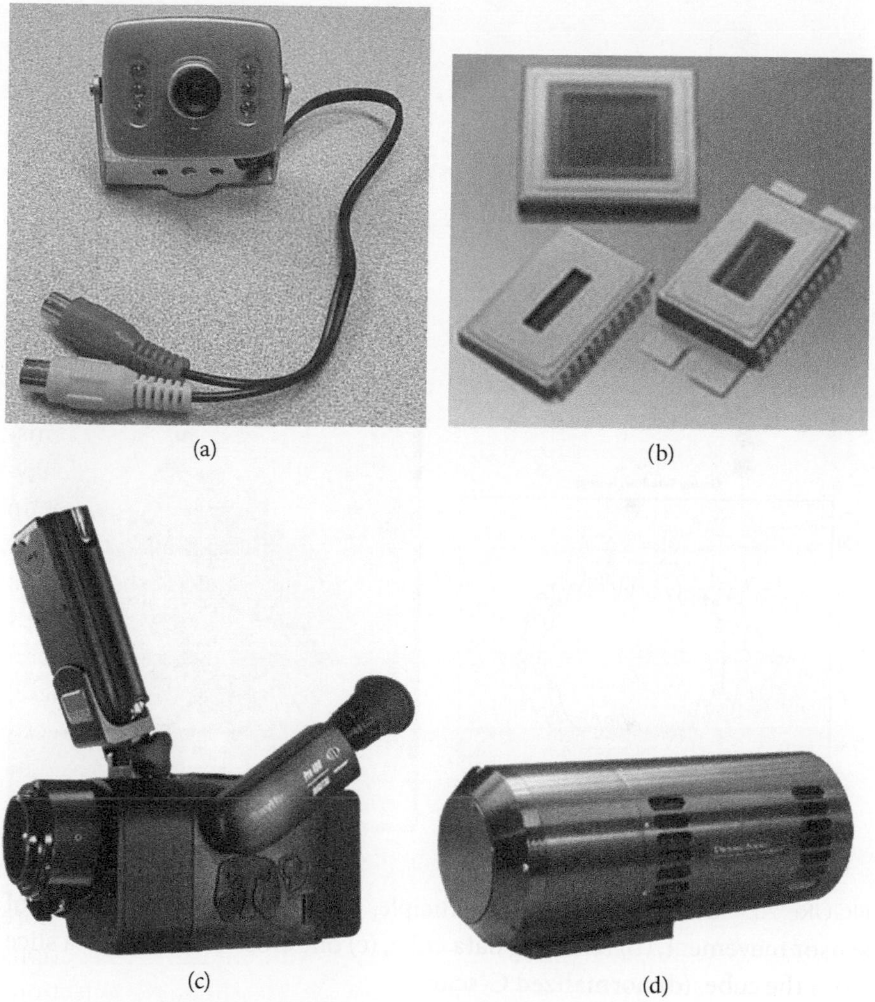

FIGURE 2.1 Direct imaging devices. (a) CCD camera, (b) CMOS photo sensor, (c) IR camera, (d) digital x-ray camera.

Similarly, MRI utilizes the changes in three external perpendicular magnetic fields passing through the patient's body due to the variations in the body part's own magnetic activities. These variations are then recorded for positions and normalized alterations in amplitude, which are then converted to images in a similar fashion as in ultrasonic imaging.

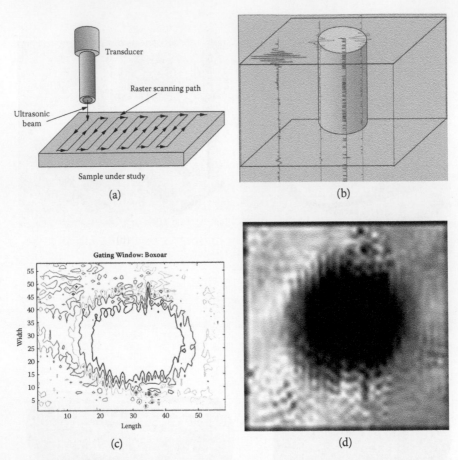

FIGURE 2.2 Ultrasonic imaging principle. (a) Raster scanning pattern of sensor movement, (b) resulting data cube, (c) one cross-sectional data slice from the cube, (d) normalized C-scan.

Another means of acquiring digital images is document scanners. A scanner is a movable camera that acquires the intensity images of the document under scanning and combines the slices to form a complete image. This procedure can be used to digitize photographs that were taken previously using regular film cameras.

In some cases, the available camera has only analog output available. In such cases, a specialized digitizing hardware, called the frame grabber, is used which captures the analog video or still images at regular sampling intervals in the form of frames of still digital images. Figure 2.3 shows these imaging systems' components.

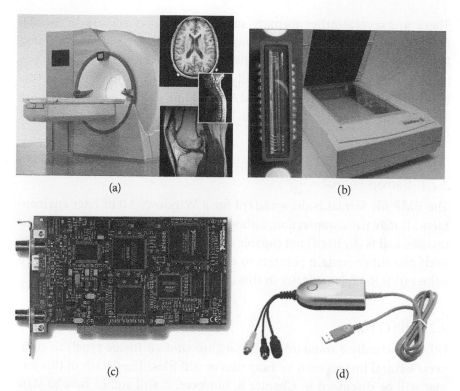

(a)   (b)

(c)   (d)

FIGURE 2.3   Imaging system components. (a) MRI scanning system with some images, (b) document scanner and its charge-coupled device (CCD) element, (c) frame grabber PCI card (NI-1411), (d) analog video to USB converter.

2.2 IMAGE TYPES AND FILE I/O

Once an image is acquired by the imaging system, it can be read into the computer's memory as a matrix. However, how this image is actually treated in the secondary storage medium varies with type of images and information content. Based on the storage types or file I/O standards, the most commonly used imaging formats in the Windows environment are as follows:

1. Bitmap (.bmp)

2. Joint Photographic Experts Group (.jpg)

3. Graphics Interchange Format (.gif)

4. Tagged Image File Format (.tif)

There are several other formats such as Windows Meta File (.wmf), Portable Network Graphics (.png), Postscript (.ps), Encapsulated Postscript (.eps), Extended Meta File (.emf), etc. However, the four types listed earlier are the most commonly used types and will be used in this book for several examples with MATLAB® files. Reading any of these types requires the function imread(); which will be discussed further in Section 2.5.

2.2.1 Bitmap Format

The BMP file format is the standard for a Windows 3.0 or later environment. It may use compression, although usually it does not compress the images, and is (by itself) not capable of storing animation. The image data itself can either contain pointers to entries in a color table or literal RGB values (this is explained later in this chapter).

2.2.2 JPEG Format

JPEG is actually a standardized algorithm (not an image type!) to compress natural images such as .bmp files or .tiff files. The details of this format will be discussed in Chapter 8; however, it will suffice here to state that this algorithm converts the natural pixels of the image using mathematical transformations and redefines the bit allocation and association structure in the image.

2.2.3 GIF Format

GIF, the brainchild of CompuServe Inc., was specifically developed for images on the Internet. It defines a protocol intended for online transmission and interchange of raster graphic data in a way that is independent of the hardware used in their creation or display. A GIF data stream is a sequence of protocol blocks and subblocks representing a collection of graphics. These blocks are connected through relational flags. GIF files are relatively easy to work with but cannot be used for high-precision color because only 8 bits are used to encode colors. Colors are encoded using local or global color tables and a built-in LZW compression algorithm. This compression algorithm is patented technology, and is currently owned by Unisys Corporation. This format encodes multiple images within a single file with embedded details of image locations on the screen, interlaced display, delay time between adjacent image frames, background color, and transition information.

2.2.4 TIFF Format

The TIFF format is controlled by a specification jointly written by Aldus Corporation and Microsoft. The format was designed to incorporate various formats with a futuristic vision. It is a complicated and platform-independent format made up of three unique data structures: Image File Header (IFH), Image File Directory (IFD), and Directory Entry (DE). The format is very similar to BMP, at least in terms of storage and usage.

Any of these formats can also be used for writing an image from MATLAB workspace to the secondary storage. The function imwrite() is used for this purpose and will be further discussed in Section 2.5.

2.3 BASICS OF COLOR IMAGES

Similar to several other achievements in the areas of science and engineering that were inspired by nature, image processing takes its inspiration from the best imaging system in nature, the human eye. The photo-sensitive area of the eye is called the *retina*, which contains two types of light-sensory cells called *rods* and *cones* (according to their shapes). Rods are responsible for producing the sensation of light intensity as well as perception of boundaries and edges in a view scene. The cones are responsible for color perception. These are of three types and are commonly called S-, L-, and M-type cells. The responses of the cone cells were experimentally recorded in the late 1970s, based on the maximum excitation of the cells with respect to a particular frequency of the incident light. Light of variable frequency would represent a particular color for a particular frequency. The responses of the three cone cells were found to be maximum for 420, 534, and 564 nm, respectively. These wavelengths (and consequently, the frequencies) correspond to blue, green, and red colors, respectively. Hence, it was discovered that human color perception is based on a triad of fundamental colors, RGB, which has since remained the most fundamental color model used in digital image processing. Figure 2.4 shows the human eye and the sensory cells along with the sensitivity curves for the RGB model.

This forms the basis of the most fundamental color model system called the RGB (Red, Green, Blue). In fact, once an image is loaded into

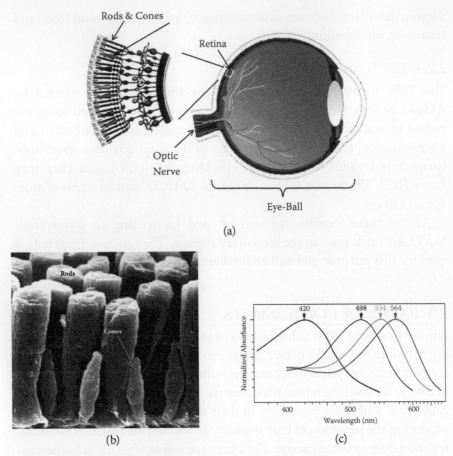

FIGURE 2.4   (See color insert following Page 204.) The human eye, its sensory cells, and RGB model sensitivity curves. (a) Diagram of the human eye showing various parts of interest, (b) cells of rods and cones under the electron microscope, (c) experimental curves for the excitation of cones for different frequencies.

a computer's memory system, most of the time it is stored as three layers of pixels corresponding to R, G, and B respectively. Figure 2.5 shows a color cube that represents various color shades based on the RGB model.

Although Figure 2.5 uses the 8-bit unsigned integer (uint8) system, the uint16, and double color cubes all have the same range of colors. In other words, the brightest red in a uint8 RGB image appears the same as the brightest red in a double RGB image. The difference is that the double RGB color cube has many more shades of red (and many more shades of all colors). Figure 2.5 shows an RGB color cube for a uint8 image.

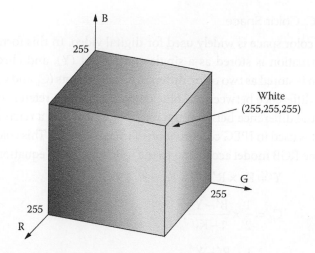

FIGURE 2.5   (See color insert following Page 204.) RGB color cube for the 8-bit unsigned integer representation system.

2.4 OTHER COLOR SPACES

MATLAB's Image Processing Toolbox·includes functions that you can use to convert RGB data to several common device-dependent color spaces, and vice versa.

2.4.1 YIQ Color Space

The National Television System Committee (NTSC) defines a color space as YIQ. This color space is used in televisions in the United States. One of the main advantages of this format is that grayscale information is separated from color data; therefore, having three components—luminance (Y), hue (I), and saturation (Q)—enables the same signal to be used for both color and black-and-white sets. *Luminance* represents grayscale information, whereas the last two components make up *chrominance* (color information). The YIQ model is related to the RGB color cube as follows:

$$\begin{bmatrix} Y \\ I \\ Q \end{bmatrix} = \begin{bmatrix} 0.299 & 0.587 & 0.144 \\ 0.596 & -0.275 & -0.321 \\ 0.212 & -0.528 & 0.311 \end{bmatrix} \begin{bmatrix} R \\ G \\ B \end{bmatrix}. \qquad (2.1)$$

$$\begin{bmatrix} R \\ G \\ B \end{bmatrix} = \begin{bmatrix} 0.9736 & 0.9873 & 0.5682 \\ 0.9675 & -0.2385 & -0.6941 \\ 0.9789 & -1.0779 & 1.6497 \end{bmatrix} \begin{bmatrix} Y \\ I \\ Q \end{bmatrix}. \qquad (2.2)$$

2.4.2 YC$_b$C$_r$ Color Space

The YC$_b$C$_r$ color space is widely used for digital video. In this format, luminance information is stored as a single component (Y), and chrominance information is stored as two color-difference components (C$_b$ and C$_r$). C$_b$ represents the difference between the blue component and a reference value. C$_r$ represents the difference between the red component and a reference value. This format is used in JPEG compression of images as well. This color space is related to the RGB model according to the following set of equations:

$$Y' = K_r \times R' + (1 - K_r - K_b) \times G' + K_b \times B',$$

$$C_b = \frac{1}{2} \times \frac{B' - Y'}{1 - K_b}, \tag{2.3}$$

$$C_r = \frac{1}{2} \times \frac{R' - Y'}{1 - K_r}.$$

Per TV standard ITU-R BT.601 (formerly known as CCIR 601), the value of K$_b$ is 0.114 and of K$_r$ is 0.299.

2.4.3 HSV Color Space

The HSV color space (hue, saturation, value) is often used by people who select colors (e.g., of paints or inks) from a color wheel or palette, because it corresponds better to how people experience color than the RGB color space does. As hue varies from 0 to 1.0, the corresponding colors vary from red through yellow, green, cyan, blue, magenta, and back to red, so

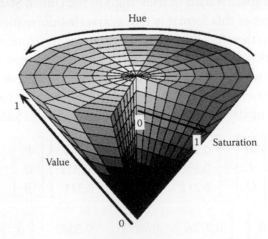

FIGURE 2.6 (See color insert following Page 204.) Illustration of the HSV color space.

that there are actually red values both at 0 and 1.0. As saturation varies from 0 to 1.0, the corresponding colors (hues) vary from unsaturated (shades of gray) to fully saturated (no white component). As value, or brightness, varies from 0 to 1.0, the corresponding colors become increasingly brighter. Figure 2.6 illustrates the HSV color space.

2.5 ALGORITHMIC ACCOUNT

In order to work with RGB images in any programming environment, basic information about size and data precision representation must be obtained before actual processing starts. An RGB image, also called *truecolor image*, of screen dimensions M rows and N columns, has a size of M × N × 3 pixels. Here, 3 represents the three layers of red, green, and blue intensities. Hence, an RGB pixel is, in fact, a 3-tuple vector with values corresponding to the red, green, and blue intensities on the screen. Graphics file formats store RGB images as 24-bit images, where the red, green, and blue components are 8 bits each. This yields a potential of 16 million colors. The precision with which a real-life image can be replicated has led to the commonly used term *truecolor image*. An RGB array can be of class double, uint8, or uint16. In an RGB array of class double, each color component is a value between 0 and 1. A pixel whose color components are (0,0,0) is displayed as black, and a pixel whose color components are (1,1,1) is displayed as white. The three color components for each pixel are stored along the third dimension of the data array. Figure 2.7 illustrates this.

The other types of image that are most commonly used are indexed images (MATLAB specific), grayscale, and binary images. Assuming the same screen dimensions as those of the foregoing RGB image, these images will respectively be M × N + P × 3, M × N, and M × N. The index images are specific to MATLAB and store the RGB colors as a separate color map of size P × 3, where P represents the total color combinations present in the RGB image. The M × N component of the index image has only the table entry index from the color map that corresponds to the color present at that particular pixel in the M × N image space. P is usually a small finite number and, therefore, there is a remarkable savings in the space requirements for storing such an image.

The grayscale image of size M × N would contain only the intensity values for that particular pixel. Conversion from RGB to grayscale depends on the sensitivity response curve of the sensor to light as a function of wavelength. However, a common equation used for this purpose is

$$gray = 0.3 \times R + 0.59 \times G + 0.11 \times B. \tag{2.4}$$

FIGURE 2.7 Anatomy of an RGB image. The highlighted area in the top image is zoomed in for the underlying values in the RGB layers and 10 × 10 blocks are shown for these layers.

(a)

(b)

(c)

(d)

FIGURE 2.8 (See color insert following Page 204.) Anatomy of image types. (a) Index image with 128 color levels showing the values of indices and the map for a selected area in the image, (b) index image with 8 color levels, (c) grayscale image for index image with 8 levels, (d) binary images with thresholding at 90%, 50%, and 10% (from left to right).

This flattens the M × N × 3 RGB image to an M × N intensity image. A binary image can be the next logical step of progression in this direction, whereas a grayscale image is thresholded according to a previously set limit. Hence, any value below this limit is made 0, and all other values are made 1. Figure 2.8 depicts these imaging structures. Obviously, the

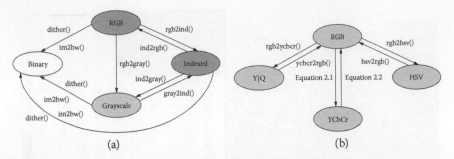

FIGURE 2.9 (a) Image-type conversions, (b) color space conversions.

grayscale image can have any value between 0 to 255, and the binary images can have only 0 or 1 for black and white areas, respectively.

2.5.1 Image Conversions

As can be deduced from the previous account, one type of image can be converted to another type. This implies that RGB can be converted to index, to grayscale, and finally to binary. Not all of these conversions are usually reversible. For instance, conversion to binary image is irreversible because the information related to individual grayscale values is lost owing to thresholding. Similarly, RGB can be converted to grayscale, but the process can be reversed if the color map is also saved for the reverse transformation. Figure 2.9(a) represents these conversions in the form of a small block diagram. Although there are MATLAB functions written next to the connecting arrows representing which function would do this transformation, the logical flow can be implemented in any programming environment by employing the appropriate specialized function from that programming environment. Similar types of conversions also exist for converting color spaces from RGB to YC_bC_r, HSV, and YIQ; this is summarized in Figure 2.9(b).

2.6 MATLAB CODE

As stated earlier, image processing begins with image acquisition. MATLAB is equipped with an acquisition toolbox called IMAQ, and has useful functions to acquire images from a Windows-recognized camera or other video source. In fact, only a handful number of statements is sufficient to do this using a simple webcam. This procedure is now described.

First, the video source is associated with a local variable by using the function videoinput(). The parameter for this function is the name of the video adapter in the Windows system. If this information is not available, then the default adapter winvideo can be used. This adapter will automatically be directed to the video source understood by Windows. More information about this particular device can be obtained through the function imaqhwinfo().

Once the device is configured, the image can be previewed using the preview() function to create a typical "viewfinder" window. The actual image can be acquired directly into a local variable using the function getsnapshot(). Multiple snapshots can be used by employing standard looping techniques. The function delete() helps in freeing some memory once the acquisition is done. The following is a typical code snippet for acquiring an image:

```
x = videoinput('winvideo',1);
imaqhwinfo(x)
preview(x)
y = getsnapshot(x);
imshow(y)
delete(x)
```

The result of the second line is the following text:

```
ans =
      AdaptorName: 'winvideo'
      DeviceName: 'Labtec webcam plus'
      MaxHeight: 288
      MaxWidth: 352
      NativeDataType: 'uint8'
      TotalSources: 1
      VendorDriverDescription: 'Windows WDM Compatible
Driver'
      VendorDriverVersion: 'DirectX 9.0'
```

The remaining lines will display two images, one for the preview and the other one for the actual acquisition allocation.

The basic data structure in MATLAB is the *array*, an ordered set of real or complex elements. This object is naturally suited to the representation of *images*, real-valued ordered sets of color or intensity data.

MATLAB stores most images as two-dimensional arrays (i.e., matrices) in which each element of the matrix corresponds to a single *pixel* in the displayed image. (Pixel is derived from *picture element* and usually denotes a single dot on a computer display.) For example, an image composed of 200 rows and 300 columns of different colored dots would be stored in MATLAB as a 200-by-300 matrix. Some images, such as RGB, require a three-dimensional array, where the first plane in the third dimension represents the red pixel intensities, the second plane represents the green pixel intensities, and the third plane represents the blue pixel intensities. Figure 2.10 shows the sample code presented in this section.

The program starts with loading the source image valley.jpg using the imread() function. Once read, the image appears as a three-dimensional array of size $426 \times 640 \times 3$, where the image has 426 rows and 640 columns and the three layers correspond to the RGB triad. How does the same image look if only one of the layers is active and the other two are all set to zero? This is achieved through the calculation of xr, xg, and xb. The original image and these components are shown in Figure 2.11. The other two layers must be set to zero in order to get true color perception; otherwise, each layer is a special case of the grayscale image.

Next, a window size of 100 by 100 is defined to highlight an area in the RGB image, and the first ten values of this area are displayed for each layer in order to more clearly understand the various combinations involved in RGB image formation. This was illustrated in Figure 2.7. The next block of processing involves the conversion of the RGB image to an index image. For this purpose, two color levels were specified as 128 and 8. Figure 2.8(a) shows the 128-level index image. This implies that the color map for this image will have 128 entries depicting 128 significant colors. This, however, is quite good in terms of approximating the image's view so that it appears the same as that of the original RGB image. However, when the index image is formed using 8 levels only, the effects of color contouring become quite obvious, as shown in Figure 2.7(b).

After that, the colored images are converted to grayscale. Again, the RGB would give an excellent grayscale equivalent, similar to the grayscale representation of the 128-level index image. However, the 8-level index image showed the contouring effect more prominently, as can be seen in Figure 2.7(c).

The grayscale images are then converted to binary images by setting three different thresholds: 90%, 50%, and 10%. As can be seen from Figure 2.7(d), 90% threshold would result mostly in dark areas because most of the grayscale values are below this limit and are therefore converted to zeroes. On the other hand, a 10% threshold is too low and, hence, most of the grayscale values are above this limit and are set to 1. In fact, an optimal thresholding is still an interesting research area, where the target is to identify the best threshold for an image in a particular imaging system.

```
clear all; close all
x = imread ('valley.jpg'); % source RGB image
xr = x;
xr (:, :, [2 3]) = 0; % initializing the Green and Blue
to zero
xg = x;
xg (:, :, [1 3]) = 0; % initializing the Red and Blue
to zero
xb = x;
xb (:, :, [1 2]) = 0; % initializing the Red and Green
to zero

N = 100; % size of the block to be isolated from x
Y = x (200:200+N, 400:400+N, :);
Y (1; 10, 1:10, :)

[z, map] = rgb2ind (x, 128); % RGB converted to Index
Image with 128 colors
[z8, map8] = rgb2ind(x, 8); % RGB to Grayscale conversion
with 8 levels

gx = rgb2gray (x); % RGB image to Grayscale conversion
gz = ind2gray (z8, map); % Index to Grayscale conversion

bx1 = im2bw (gx, 0.9); % Binary image with 90% threshold
bx2 = im2bw (gx, 0.5); % Binary image with 50% threshold
bx3 = im2bw (gx, 0.1); % Binary image with 10% threshold

figure ; imshow (x)
figure ; imshow (y)
figure ; imshow (z, map)
figure ; imshow (xr)
figure ; imshow (xg)
figure ; imshow (xb)
figure ; imshow (z8, map)
figure ; imshow (gx)
figure ; imshow (gz)
figure ; imshow (bx1)
figure ; imshow (bx2)
figure ; imshow (bx3)
```

FIGURE 2.10 Sample MATLAB code for various image conversion operations.

(a)   (b)

(c)   (d)

FIGURE 2.11   (See color insert following Page 204.) Original RGB image with its components. (a) Source RGB image, (b) red component, (c) green component, (d) blue component.

In case any of these images needs to be saved to the hard disk, the following syntax can be used:

```
imwrite(x,'True-Valley.jpg','jpg');
```

2.7 SUMMARY OF IMAGE TYPES AND NUMERIC CLASSES

In this chapter, the following important points have been covered:

- Image processing begins with image acquisition, which can be done from a number of sources such as cameras, ultrasonic images, MRI, and radiographs. The image quality and characteristics depend on the imaging system and its application.

- IMAQ toolbox provides very easy procedures to acquire images or video streams from the connected hardware.

- Images in MATLAB can be one of the following types: RGB, index, grayscale, binary, etc.

- Color spaces are interchangeable in MATLAB from RGB to HSV or YC_bC_r or YIQ. Other specialized formats are also provided in the toolbox, to which the image can be converted from RGB.

- An RGB image is composed of three layers corresponding to the red, green, and blue intensities in the image. A simple procedure has been outlined to view these color layers separately.

- Indexed images are specialized images used within MATLAB only, and can be employed for better storage utilization.

- Index images can also be used to limit the number of color levels by specifying them.

- The grayscale image is the most commonly used image form for most image processing algorithms.

- Grayscale images can have any value between 0 and 255 for unsigned integer representation.

- Binary images are formed by thresholding the grayscale image such that any value below this threshold becomes 0 and other values become 1.

- For certain operations, it is helpful to convert an image to a different image type. For example, if you want to filter a color image that is stored as an indexed image, you should first convert it to RGB format. When you apply the filter to the RGB image, MATLAB filters the intensity values in the image, as appropriate. If you attempt to filter the indexed image, MATLAB simply applies the filter to the indices in the indexed image matrix, and the results might not be meaningful.

- The following table lists all the image conversion functions in the Image Processing Toolbox.

Function	Description
dither	Create a binary image from a grayscale intensity image by dithering; create an indexed image from an RGB image by dithering
gray2ind	Create an indexed image from a grayscale intensity image
im2bw	Create a binary image from an intensity image, indexed image, or RGB image, based on a luminance threshold
ind2gray	Create a grayscale intensity image from an indexed image
ind2rgb	Create an RGB image from an indexed image
rgb2gray	Create a grayscale intensity image from an RGB image
rgb2ind	Create an indexed image from an RGB image

2.8 EXERCISES

1. Use the image file shapes.jpg, which contains four basic shapes in various colors. Analyze the image to build a map of your own with the RGB values corresponding to the colors present in the image. Then convert the original image into grayscale image so that all the color information is lost. Now apply the reverse mapping to convert it from grayscale to RGB directly. Note that all the conversions are directly from RGB to grayscale or from grayscale to RGB. No other transformation is allowed.

2. Write an m-file to use the webcam for acquiring images every fifth second, and display it as a binary image with 50% threshold.

3. Write an m-file to read a colored image. It will be read as an RGB image. Using the equation $y = 0.299\,R + 0.59\,G + 0.11\,B$, convert this RGB image into grayscale (y) ranging from 0 to 255. You may have to scale and round the values because the grayscale image values should be represented by 8-bit integers only. Your function must have this format: $y = myrgb2gray(x)$ where x is the RGB image.

4. Convert the RGB image from Question #3 into an indexed image without using the rgb2ind() function. In this exercise, you must program a function of your own to do the conversion by following the steps discussed in context of index images:

   1. Make a table of all color values as map.

   2. Assign index values of the map to the pixel position where this color occurred in the image.

   3. Your function must have this format: [y,map] = myrgb2ind(x), where x is the RGB image.

5. Convert the RGB image from Question #3 into binary by using your own thresholding program. Your function must look like bx = myim2bw(gx,t), where gx is the grayscale image to be converted to binary and t is the threshold.

CHAPTER 3

Image Arithmetic

3.1 INTRODUCTION

ONCE IMAGES ARE INSIDE the computer system, or more specifically, once they are read inside a program, the images are nothing but matrices. Hence, all the operations that can be applied to matrices should theoretically be applicable to the images as well. However, that is not entirely true. Some of the matrix-arithmetic operators can be applied to image matrices, whereas other operators do not have any meaning in image processing. At the same time, the image matrices have certain special operators that are specific to images and that are not usually found in the matrix-arithmetic world. Image arithmetic is the implementation of standard arithmetic operations, such as addition, subtraction, multiplication, and division, for images. Image arithmetic has many uses in image processing, both as a preliminary step in more complex operations and by itself.

This chapter covers some of the most commonly used operations applied to digital images. The differences from matrix arithmetic will be pointed out at the appropriate places in the discussion.

3.2 OPERATOR BASICS

Image arithmetic operators operate on either two images of exactly the same size or an image and a number. Essentially, it corresponds to two matrices of the same order or a matrix and a scalar. For the first case, the operation is performed on a pixel-by-pixel basis where each pixel corresponds to one of the elements in the matrices. For the second case, the scalar number is operated over the entire matrix on each element, that is, all the pixels. The two behaviors are shown in Figure 3.1.

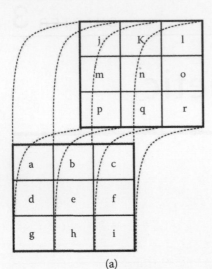
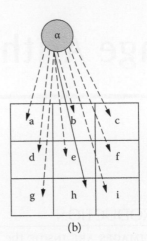

(a)   (b)

FIGURE 3.1 Two modes of image arithmetic operations. (a) Pixel-by-pixel operation between two images of same size, (b) scalar operation on an image.

3.3 THEORETICAL TREATMENT

Usually, the following image operators are used as the building blocks for complex processing:

1. Image addition

2. Image subtraction

3. Image multiplication

4. Image division

5. Image blending

The following sections will discuss these operators theoretically.

3.3.1 Pixel Addition

This operator takes as input two images of the same size and produces as output a third image of the same size as the first two, in which each pixel value is the sum of the values of the corresponding pixel from each of the two input images. The same strategy can be applied to more than two images as a continuation of the same operation. A scalar variant of the operator also exists, in which a specified constant number is added to every pixel. The addition of two images is performed straightforwardly in a single pass. The output pixel values are given by

$$R(m,n) = P(m,n) + Q(m,n), \tag{3.1}$$

where P, Q, and R represent image matrices of the same size, and m and n represent the mth and nth pixels.

Alternatively, if it is simply desired to add a constant value C to a single image, then:

$$R(m,n) = P(m,n) + C. \tag{3.2}$$

If each pixel value corresponds to an n-tuple instead of a number (for instance, the color image would have each pixel as a vector of three values corresponding to red, blue, and green components), then such pixels are operated on by addition at each corresponding vector element to produce the output value.

The results of addition can easily grow beyond the maximum value that the pixel bit resolution will allow. For 8-bit representation of pixels, the addition of two pixels with values of 200 each would result in 400, which is higher than the maximum representable number with 8 bits, 255. The overflowing pixel values are usually set to the maximum allowed value, an effect known as *saturation*.

3.3.2 Pixel Subtraction

The pixel subtraction operator takes two images as input and produces as output a third image whose pixel values are simply those of the first image minus the corresponding pixel values from the second image. It is also often possible to just use a single image as input and subtract a constant value from all the pixels. Some versions of the operator will output the absolute difference between pixel values rather than the signed output. The subtraction of two images is performed straightforwardly in a single pass. The output pixel values are given by

$$R(m,n) = P(m,n) - Q(m,n). \tag{3.3}$$

Or if the operator computes absolute differences between the two input images, then:

$$R(m,n) = |P(m,n) - Q(m,n)|. \tag{3.4}$$

Or if it is simply desired to subtract a constant value C from a single image, then:

$$R(m,n) = P(m,n) - C. \tag{3.5}$$

If each pixel value corresponds to an *n*-tuple instead of a number (for instance, the color image would have each pixel as a vector of three values corresponding to red, blue, and green components), then such pixels are operated on by subtraction at each corresponding vector element to produce the output value.

If the operator calculates absolute differences and the two input images use the same pixel value type, then it is impossible for the output pixel values to be outside the range that may be represented by the input pixel type, and so this problem does not arise. This is one good reason to use absolute differences.

3.3.3 Pixel Multiplication and Scaling

Although images are treated as matrices, image multiplication does not follow the matrix multiplication rule. This implies that instead of having the number of columns of the first matrix be the same as the number of rows in the second matrix, image multiplication requires the two images to be of the same size. Similar to other image arithmetic operators, multiplication comes in two main forms. The first form takes two input images and produces an output image in which the pixel values are just those of the first image, multiplied by the values of the corresponding values in the second image. The second form takes a single input image and produces output in which each pixel value is multiplied by a specified constant. This latter form is probably more widely used and is generally called *scaling*. The multiplication of two images is performed in the obvious way in a single pass using the formula:

$$R(m,n) = P(m,n) \times Q(m,n). \tag{3.6}$$

Scaling by a constant is performed using

$$R(m,n) = P(m,n) \times C. \tag{3.7}$$

Note that the constant is often a floating-point number, and may be less than one, which will reduce the image intensities. It may even be negative if the image format supports that.

If each pixel value corresponds to an *n*-tuple instead of a number (for instance, the color image would have each pixel as a vector of three values corresponding to red, blue, and green components), such pixels are operated on by multiplication at each corresponding vector element to produce the output value. If the output values are calculated to be larger than the maximum allowed pixel value, then they are usually truncated at that maximum value. The same is also true for values less than the minimum allowed pixel values, in which case the truncation is done at the minimum value.

3.3.4 Pixel Division

The image division operator normally takes two images as input and produces a third image whose pixel values are the pixel values of the first image divided by the corresponding pixel values of the second image. Many implementations can also be used with just a single input image, in which case every pixel value in that image is divided by a specified constant.

The division of two images is performed in the obvious way, in a single pass using the formula:

$$R(m,n) = P(m,n) \div Q(m,n). \tag{3.8}$$

Division by a constant is performed using

$$R(m,n) = P(m,n) \div C. \tag{3.9}$$

If each pixel value corresponds to an n-tuple instead of a number (for instance, the color image would have each pixel as a vector of three values corresponding to red, blue, and green components), such pixels are operated on by division at each corresponding vector element to produce the output value.

The division operator may only implement integer division, or it may also be able to handle floating-point division. If only integer division is performed, then results are typically rounded down to the next lowest integer for output. The ability to use images with pixel value types other than 8-bit integers comes in very handy when doing division.

3.3.5 Blending

This operator produces a blend of two input images of the same size by combining scalar multiplication and image addition. Similar to pixel addition, the value of each pixel in the output image is a linear combination of the corresponding pixel values in the input images. The coefficients of the linear combination are user specified, and they define the ratio by which the images must be scaled before combining them. These proportions are applied such that the output pixel values do not exceed the maximum pixel value. The resulting image is calculated using the following formula:

$$R(m,n) = C \times P(m,n) + (1-C) \times Q(m,n). \tag{3.10}$$

P and Q are the two input images. C is the blending ratio, which determines the influence of each input image in the output. C can either be a constant factor for all pixels in the image or can be determined for each pixel separately using a mask. The size of the mask must be identical with the size of the images.

3.4 ALGORITHMIC TREATMENT

In this section, details related to algorithmic structures and usage of the operators mentioned in Section 3.3 will be presented. The emphasis is on presenting the algorithms in a generic manner so that they can be programmed in any language.

3.4.1 Image Addition

Figure 3.2(a) represents the algorithmic structure for adding two images. Obviously, the operation is repetitive and can be realized using any recursive programming approach, including loops, or a recursive function call. Image addition is usually a supportive operation rather than an operation on its own in image processing. For example, a constant value can be added to the image to brighten up the view, or enhanced edges can be added in an image to sharpen the object linings and new images can be generated through addition of two or more other images.

One important point to be mentioned here is the fact that although the theory presented or the equations mentioned all deal with two images or an image with a scalar, the same approach can be extended to more than two images by repeating the algorithms as needed.

3.4.2 Image Subtraction/Multiplication/Division

Once the operation for addition is understood from Section 3.4.1, the remaining three operations, subtraction, multiplication, and division, can be considered similarly. As can be seen from Figures 3.2(b), (c), and (d), the algorithms are exactly the same as for addition except for the main processing operation, which is changed according to the desired operations. Hence, the calculation requirements change at the pixel level only, which also depends on the processor in use and the way in which these operators are implemented at the hardware level.

Image subtraction can be used as a preliminary step in more complex image processing, or by itself. For example, one can use image subtraction to detect changes in a series of images of the same scene, or subtraction can be used to separate the foreground artifacts from background for applications such as target detection, scene analysis, localized processing, and

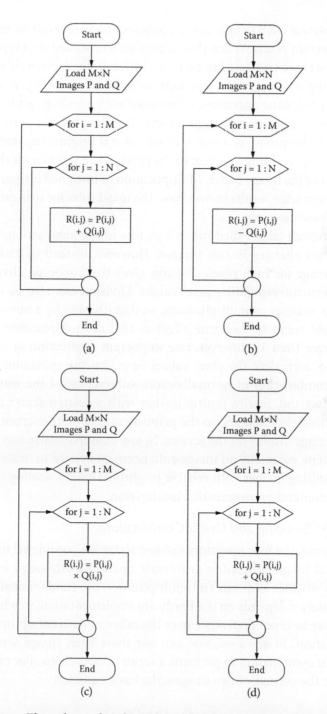

FIGURE 3.2 Flow charts for algorithmic implementation of (a) image addition, (b) image subtraction, (c) image multiplication, (d) image division.

region of interest (ROI) applications. Subtraction can result in negative values for certain pixels. When this occurs with unsigned data types, the negative values are truncated to zero (0), which is displayed as black.

Multiplying an image by a constant, referred to as *scaling*, is a common image processing operation. When used with a scaling factor greater than 1, scaling brightens an image; a factor less than 1 darkens an image. Scaling generally produces a much more natural brightening/darkening effect than simply adding an offset to the pixels since it preserves the relative contrast of the image better. Multiplication of unsigned integer values in images very often results in overflow. The usual rules for truncation are applied in these cases, too.

Image division (also called *rationing*), like image subtraction, can be used to detect changes in two images. However, instead of giving the absolute change for each pixel, division gives the fractional change, or ratio, between corresponding pixel values. Division can also be used in the opposite manner as multiplication, so that division by a number less than 1 would result in the same effect as that of multiplication with a number larger than 1. However, one important application of division would be to normalize the pixel values based on the maximum value. In fact, a combination of normalization, subtraction of the minimum value as offset and, finally, multiplication with a desired scalar to scale the normalized shifted pixels to the printing ranges on the screen is very useful for image display on the screen. In any comprehensive image processing system, especially an image enhancement system in order to display the resulting images with relative brightness values, scaling must be done for better and more realistic visualization.

3.4.3 Image Blending and Linear Combinations

To some extent, the four operators defined so far are considered to be the fundamental image processing arithmetic operators. Although one may argue as to whether addition and multiplication are fundamental operators, ultimately it depends on the hardware implementation, in which one operator may be given preference over the other in terms of optimality of implementation. In any case, you can use these basic image arithmetic functions in combination to perform a series of operations. For example, to calculate the average of two images, the basic equation

$$R(i,j) = \frac{P(i,j) + Q(i,j)}{2},$$
(3.11)

can be implemented as any of the following:

1. Adding images P and Q first, and then dividing the sum by 2

2. Dividing both P and Q by 2, and then adding them

3. Multiplying P and Q by 0.5, and then adding them

4. Adding P and Q, and then multiplying by 0.5, etc. ...

Hence, it will depend on the hardware implementation, which may favor one operation over the other. Image blending, in a similar way, is another classical example of a compound operator resulting from a linear combination of the fundamental operators. Figure 3.3 shows the algorithmic implementation of the image blending operation.

A typical application of the blending operation is in image generation, in which two different images can be combined to give special effects of transparency, fading, or added objects in the original image.

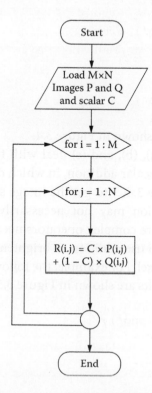

FIGURE 3.3 Algorithm for image blending operation.

3.5 CODING EXAMPLES

In this section, the various arithmetic operators discussed so far will be explained with examples of codes in MATLAB® using its Image Processing Toolbox. The given examples are related to color images and, hence, the 3-tuple treatment is performed on the images. Once read inside the computer, the images are in unsigned integer format of 8 bits (uint8) and, hence, the toolbox functions are used for converting the types and taking care of truncation, etc.

In the following examples, some of the sample images are used for convenience in presenting the main point. Although simple, these images do correspond well with the real-world scenario and can be transparently applied in that case.

3.5.1 Image Addition

Consider the following code:

```
P = imread('sml1.jpg') ;
imshow(P)
Q = imread('sml2.jpg') ;
figure ; imshow(Q)
R = imadd(P,Q) ;
figure ; imshow(R)
figure ; imshow(R+100)
```

The resulting images are shown in Figure 3.4.

Whereas Figures 3.4(a), (b), and (c) deal with the foregoing code, the fourth image represents scalar addition, in which case 100 is added to the resulting image of Figure 3.4(c) to get the image shown in Figure 3.4(d). Obviously, simple addition may not necessarily produce the desired results, and probably more complex operators need to be built to suit the application. You can also use addition to brighten an image by adding a constant value to each pixel. For example, the following code brightens an RGB image, and the results are shown in Figure 3.5.

```
P = imread('peppers.png');
Rs = imadd(P, 100);
figure ; imshow(P);
figure ; imshow(Rs);
```

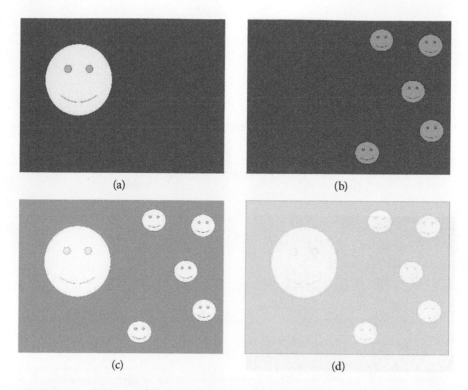

FIGURE 3.4   Image addition example. (a) Original image P, (b) original image Q, (c) resulting image $R = P + Q$, (d) resulting image with scaling $R + 100$.

FIGURE 3.5   Brightening an RGB image by adding a constant value to each pixel. (a) Original image, (b) resulting image after adding 100 to the original image.

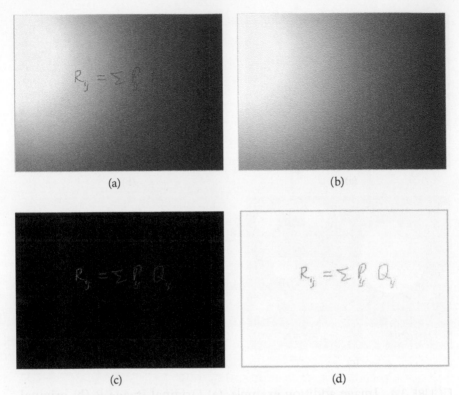

(a) (b)

(c) (d)

FIGURE 3.6 Example of using image subtraction. (a) Original image from a whiteboard, (b) background image, (c) result of absolute difference, (d) result of complementing the image in (c) by subtracting it from 255.

3.5.2 Image Subtraction

Image subtraction is used both as a substep in complicated image processing sequences, and also as an important operation in its own right. Two examples of the usage of this operator follow.

A common use is to subtract background variations in illumination from a scene so that the foreground objects in it may be more easily analyzed. Figure 3.6(a) shows a whiteboard image taken in very poor illumination conditions. Figure 3.6(b) is another snapshot of the same whiteboard without any text on it, which is then subtracted from the actual image. The resulting difference is calculated using Equation 3.4. However, the unwanted black color is due to the difference of the same-valued pixels. The subtraction operator can be applied again by subtracting this resulting image from 255, which is the maximum value of a pixel for 8-bit representation. This results in a complementary image for the previous result that is more visually suitable than the previous one. These last two images are shown in Figures 3.6(c) and (d), and the underlying code is as follows:

```
P = imread('back1.jpg') ;
imshow(P)
Q = imread('back2.jpg') ;
figure ; imshow(Q)
R = imabsdiff(P,Q) ;
figure ; imshow(R)
Rc = 255 - R ;
figure ; imshow(Rc)
```

A key application of image subtraction is in detecting the changes in a scene, an application used in target tracking and movement detection in surveillance applications. Figures 3.7(a) and (b) show two scenes that differ very minutely. The difference is so small that it is hard for a human user to detect the changes. However, the application of absolute image subtraction results in the changes being revealed very prominently, as shown in Figure 3.7(c). A complementary view in Figure 3.7(d) reveals the changes even more prominently.

The following is the underlying code that generated the foregoing images:

```
P = imread('chng1.jpg') ;
figure ; imshow(P)
Q = imread('chng2.jpg') ;
figure ; imshow(Q)
R = imabsdiff(P,Q) ;
figure ; imshow(R)
Rc = 255 - R ;
figure ; imshow(Rc)
```

3.5.3 Multiplying Images

To multiply two images, use the immultiply function, which performs an element-by-element multiplication (.*) of corresponding pixels in a pair of input images and returns the product of these multiplications in the corresponding pixel in an output image. However, there are very limited applications for such a pixel-by-pixel implementation. For example, the filtering operation in the frequency domain is done by first taking the Fourier transforms of the noisy image as well as the filter and then multiplying them pixel by pixel, after which an inverse Fourier transform is taken for the product. This example will be covered in Chapter 6, which deals with image filtering.

However, the most common application of image multiplication is to scale down the image values so that they can be within the two saturation

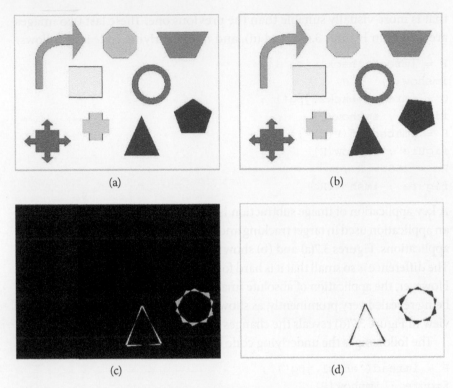

(a)　　　　　　　　　　　　(b)

(c)　　　　　　　　　　　　(d)

FIGURE 3.7 Detecting changes in the scene using image subtraction. (a) Original image, (b) image with changes, (c) changes detected with `imabsdiff` function, (d) complement image of (c).

limits (minimum and maximum) for appropriate viewing. The need for such an operation arises on two occasions: first, if the source image is too dark or too bright, and second, as a result of some processing step, the resulting matrix has values that are very high. A typical example of the second situation is convolution of two images, where products of sums can result in very high values. This is realized in the following:

```
P = imread('coins.png') ;
figure ; imshow(P)
Q = fspecial('unsharp',0.5) ;
figure ; mesh(Q)
R = conv2(P,Q) ;
figure ; imshow(R)
figure ; imshow(R*0.005)
```

Figure 3.8 shows the results of this code.

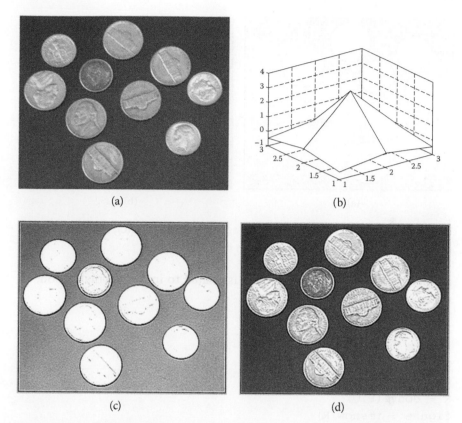

(a) (b)

(c) (d)

FIGURE 3.8 Scaling down image values by image multiplication. (a) Original image, which is very soft and can be sharpened, (b) the sharpening filter used, (c) resulting image of the convolution without scaling, (d) scaled image from convolution.

3.5.4 Dividing Images

Image division can be used the same way as image multiplication for scaling purposes. As such, the code line

```
imshow(R*0.005);
```

can be written as

```
imshow(R/200);
```

to give the same results as in Figure 3.8.

To divide two images, use the imdivide function. The imdivide function performs an element-by-element division (./) of corresponding pixels in a pair of input images. The immultiply function returns the result in the corresponding pixel in an output image. However, an important

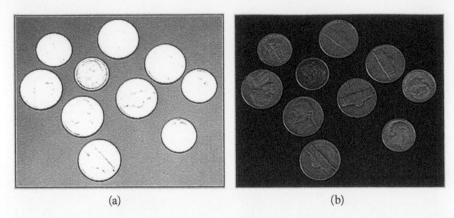

(a) (b)

FIGURE 3.9   Image division. (a) is the same as Figure 3.8(c); (b) is the resulting image after normalizing through image division.

application of the division operator is related to image normalization, as shown in the following code:

```
P = imread('coins.png') ;
figure ; imshow(P)
Q = fspecial('unsharp',0.5) ;
figure ; mesh(Q)
R = conv2(P,Q) ;
figure ; imshow(R)
figure ; imshow(R/max(max(R)))
```

The results are shown in Figure 3.9.

3.5.5 Image Blending and Linear Combinations

Image blending is implemented in MATLAB either using the empirical formula of Equation 3.10, or by using the function imlincomb. When used with uint8 or uint16 data, each arithmetic function truncates its result before passing it on to the next operation. This truncation can significantly reduce the amount of information in the output image. In fact, a better way to perform this series of calculations is to use the imlincomb function instead of the empirical formulation. imlincomb performs all the arithmetic operations in the linear combination in double precision and only truncates the final result. The following code shows how this function can produce a merged image from two images:

```
P = imread('circles.png') ;
figure ; imshow(P)
```

```
Q = imread('testpat1.png') ;
figure ; imshow(Q)
x = 0.8 ;
R = imlincomb(x,P,1-x,Q);
figure ; imshow(R)
x = 0.2 ;
R = imlincomb (x,P,1-x,Q);
figure ; imshow (R)
```

The results from this code are shown in Figure 3.10.

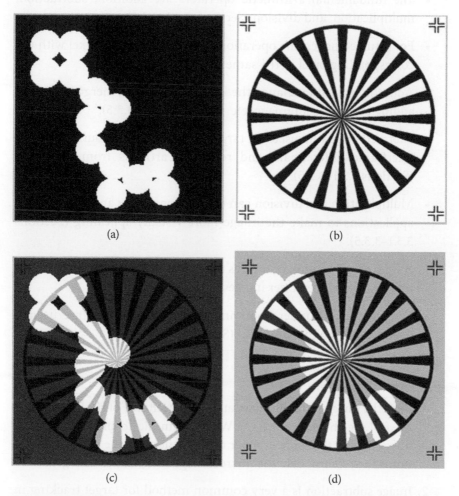

(a)

(b)

(c)

(d)

FIGURE 3.10 Results of code merging two images. (a) Original image P, (b) original image Q, (c) and (d) are results of $xP + (1 - x)Q$ for $x = 0.8$ and 0.2, respectively.

3.6 SUMMARY

- Arithmetic operators can operate on images in much the same way as matrices, except for a few considerations (Section 3.1).

- Image arithmetic works in a pixel-by-pixel manner for two images of the same size (Section 3.2).

- The other form of image arithmetic is to apply a scalar over the image matrix (Section 3.2).

- The fundamental arithmetic operators are addition, subtraction, multiplication, and division (Sections 3.3.1–3.3.5).

- For *n*-tuple images, the operation takes place for each pixel with the corresponding pixel in the same layer (Section 3.3.3).

- Addition is used to change the intensity of pixels, merging images, adding content to an image, etc. (Sections 3.3.1–3.3.5).

- Subtraction can be used for target detection, background removal, interframe differentials, and removing artifacts from an image (Sections 3.3.1–3.3.5).

- Multiplication and division can be used for scaling and normalizing applications to make the image more visually appealing (Sections 3.3.1–3.3.5).

- A linear combination of images is also possible and is used to merge images to produce newer images (Sections 3.3.1–3.3.5).

- Overshoots and undershoots in terms of pixel values are usually dealt with through truncation (Section 3.3.3).

3.7 EXERCISES

1. Read the image of an object located on the left side of the image [such as sml1.jpg in Figure 3.4(a)]. Write an m-file to produce another image having the same number of rows but twice the number of columns to accommodate the mirror image of the source image.

2. Image subtraction is a very common method for target tracking in an image. Use several successive images from a video stream, and track a particular target in it using this method.

3. Write an m-file to read the images img1.jpg, and img2.jpg. Use the cross-correlation function from MATLAB between img1 and img2 with appropriate thresholding to detect and count the circles in img1.jpg.

4. Repeat Question #3 with the convolution operation instead of cross-correlation. Is the result similar to the previous one? Comment.

5. Use two grayscale images x and y, and prove through MATLAB operations that $x.^*y = \log^{-1}[\log(x) + \log(y)]$. Make sure that the images do not have any zero value before attempting to take logarithms.

CHAPTER 4

Affine and Logical Operations, Distortions, and Noise in Images

4.1 INTRODUCTION

IN MANY IMAGING SYSTEMS, detected images are subject to geometric distortion introduced by perspective irregularities in which the position of the camera with respect to the scene alters the apparent dimensions of the scene geometry. Applying an affine transformation to a uniformly distorted image can correct for a range of perspective distortions by transforming the measurements from the ideal coordinates to those actually used. Similarly, there are situations related to masking out certain areas of an image. This is done through logical operations that act on the image in a pixel-by-pixel manner and inhibit or manifest certain values depending on the operation. In this chapter, these two types of operations are explored in detail so that the reader can prepare for the advanced operations where these two operations will be used repeatedly to perform a higher level task.

4.2 AFFINE OPERATIONS

An affine operation or transformation is an important class of linear 2D geometric transformations that maps variables [e.g., pixel intensity values located at position (x_1, y_1) in an input image] into new variables [e.g., (x_2, y_2) in an output image] by applying a linear combination of translation, rotation, scaling, and/or shearing (i.e., nonuniform scaling in some directions) operations. Usually referred to as TRS operators, these affine operators are extensions of the linear operator with an added bias. This greatly helps a machine vision system to correct for perspective errors; that is, an object may appear different if it is not straight but tilted slightly. The affine operations can correct for the perspective errors by rotating, shifting, and scaling certain portions of the image.

The general affine operation is commonly written in homogeneous coordinates as follows:

$$\begin{bmatrix} x_2 \\ y_2 \end{bmatrix} = A \begin{bmatrix} x_1 \\ y_1 \end{bmatrix} + B. \tag{4.1}$$

The TRS operations are summarized in Table 4.1. Several different affine transformations are often combined to produce a resultant transformation. The order in which the transformations occur is significant because a translation followed by a rotation is not necessarily equivalent to the converse.

Figure 4.1 depicts some of the affine operations on a sample image. In these images, the original image (a) goes through the TRS transforms. In Figure 4.1(b), the image appears shuffled because each of its quarter portions has been shifted (translated) to the location of another portion. For instance, portions 1 and 4 are swapped while 2 and 3 are swapped. The swapping is done through adding or subtracting the appropriate displacement from or to each pixel of the portion. Figure 4.1(c) shows the rotation by 45° (counterclockwise rotation). The resulting black portions are zero-padding regions to keep the overall image rectangular or square in shape. It is interesting to note that the image size increases due to this operation. For this particular case, the size increases from 480 × 640 × 3 to 793 × 793 × 3. If an equal rotation is performed in the opposite direction, the image will appear straight but the size will increase to 1125 × 1125 × 3. The last image in Figure 4.1 shows the effect of scaling, where the image size reduces to its quarter by replacing each group of $a_{11} \times a_{22}$ pixels with its average value. This can be clearly seen through the size of the resulting image as well as the blurriness produced by this operation.

Most implementations of the affine operator allow the user to define a transformation by specifying where three (or less) coordinate pairs from the input image (x_1, y_1) remap in the output image (x_2, y_2). Once the

TABLE 4.1   Affine TRS Operations

Operation	Operators
Translation: Pixel movement by b_1 in x and b_2 in y direction. The treatment of elements near image edges varies with implementation. Translation is used to improve visualization of an image, but also has a role as a preprocessor in applications where registration of two or more images is required.	$A = \begin{bmatrix} 1 & 0 \\ 0 & 1 \end{bmatrix}$, $B = \begin{bmatrix} b_1 \\ b_2 \end{bmatrix}$
Rotation: Rotating all pixels by an angle of θ degrees (counterclockwise for positive angle). In most implementations, output locations (x_2, y_2) that are outside the boundary of the image are ignored. Rotation is most commonly used to improve the visual appearance of an image, although it can be useful as a preprocessor in applications where directional operators are involved.	$A = \begin{bmatrix} \cos\theta & -\sin\theta \\ \sin\theta & \cos\theta \end{bmatrix}$, $B = \begin{bmatrix} x_0 \\ y_0 \end{bmatrix}$ x_0 and y_0 are the origin's coordinates.
Scaling: Performs a geometric transformation that can be used to shrink or zoom the size of an image. Image reduction, commonly known as *subsampling*, is performed by replacement (of a group of pixel values by one arbitrarily chosen pixel, a_{11} or a_{22}, value from within this group) or by *interpolating* between pixel values in a local neighborhood. Image zooming is achieved by *pixel replication* or by interpolation.	$A = \begin{bmatrix} a_{11} & 0 \\ 0 & a_{22} \end{bmatrix}$, $B = \begin{bmatrix} 0 \\ 0 \end{bmatrix}$ a_{ii} represents the number of pixels in a selected block per adopted methodology.

transformation has been defined in this way, the remapping proceeds by calculating, for each output pixel location (x_2, y_2), the corresponding input coordinates (x_1, y_1). If that input point is outside of the image, then the output pixel is set to the background value. Otherwise, the value of (1) the input pixel itself, (2) the neighbor nearest to the desired pixel position, or (3) a bilinear interpolation of the neighboring four pixels is used.

Affine operations are most commonly applied when an image is detected to have undergone some type of distortion. The geometrically correct version of the input image can be obtained from the affine transformation by resampling the input image such that the information (or intensity) at each point (x_1, y_1) is mapped to the correct position (x_2, y_2) in a corresponding

FIGURE 4.1   (See color insert following Page 204.) Various affine operations. (a) Original image, (b) translated by (±240, ±320), (c) rotated by 45°, (d) scaled by 25%.

output image. These operations assist in image registration, correction for perspective distortions in the image, and alignment in terms of angle as well as size for later processing.

4.3 LOGICAL OPERATORS

The affine or geometric operations affect the image coordinate values, but sometimes there is a need to highlight or completely remove some parts of the image. This usually happens when a specific portion of an image needs to be removed for certain machine vision applications. In general, the logical operations are binary in nature and most of the time operate on binary images as well, but they can be tuned to work with grayscale

TABLE 4.2   Logical Operators for Images

Operation	Binary operator				Grayscale operator
AND and NAND: The most obvious application of AND is to compute the intersection of two images. It is also used to extract a portion of an image.	A	B	A&B	(A&B)	$A\&B = \min(A,B)$ Or $AB = \min(A,B)$
	0	0	0	1	
	0	1	0	1	
	1	0	0	1	
	1	1	1	0	
OR and NOR: The OR operator typically outputs a third image whose pixel values are just those of the first image ORed with the corresponding pixels from the second.	A	B	A+B	(A+B)	$A+B = \max(A,B)$
	0	0	0	1	
	0	1	1	0	
	1	0	1	0	
	1	1	0	1	
XOR and XNOR: For two inputs, the XOR function is only true if just one (and only one) of the input values is true, and false otherwise.	A	B	A⊕B	(A⊕B)	$A \oplus B = AB' + A'B$
	0	0	0	1	
	0	1	1	0	
	1	0	1	0	
	1	1	0	1	
NOT: *Logical NOT*, or *invert*, is an operator that takes a binary or gray-level image as input and produces its photographic negative; that is, dark areas in the input image become light and light areas become dark.	A		A'		$A' = 2^b - 1 - A$ Where b is the number of bits representing the grayscales
	0		1		
	1		0		

images. In essence, the operators work on the same principles as discrete gates in digital logic circuit designs. Table 4.2 summarizes the basic logical operations commonly used in image processing.

Figure 4.2 shows these operations in action. Two grayscale images, A and B, are used. B is a very simple image and has been selected because it can show the effects of logical operations in a much clearer manner than a general image. Figures 4.2(c) and (d) show the logical inverts of the two source images. Such inverts are also commonly called *negatives* of the original image. By ANDing the two images, the dominant area of the masking image, B, reveals a specific region of A and hence can be used to isolate or mask out certain parts of an image. The next image is the invert of the ANDed image and shows the negative effect for the previously masked

FIGURE 4.2 Demonstration of logical operations. (a) Original image A, (b) original image B, (c) A', (d) B', (e) AB, (f) (AB)', (g) A + B, (h) (A + B)', (i) A⊕B, (j) (A⊕B)'.

(g)

(h)

(i)

(j)

FIGURE 4.2 (Continued)

area. The next two images are the result of OR and NOR operations, which can be used to remove a specific area from the image or to isolate the masked area. The last two images represent the XOR and XNOR operations. As can be seen from Figures 4.2(i) and (j), these operations provide better isolation of the masked regions and can be used to generate the artificial binocular effect, as if a specific area is the target of a specialized binocular vision that has highlighted the image in that portion only.

4.4 NOISE IN IMAGES

In a machine vision system, images acquired through modern sensors may be contaminated by a variety of noise sources. This phenomenon is not very visible for the usual photographic cameras due to a very stable electronics environment and very short imaging paths, which are not the usual conditions in the machine vision system or in extreme applications such as military, satellite, and highly noisy vibrating environments in the industry. Noise refers to the stochastic variations in the image pixel values. In most of the imaging applications where photo imaging is of concern, charge-coupled device (CCD) cameras are used, where photons produce electrons that are commonly referred to as photoelectrons. Nevertheless, most of the observations outlined in the following sections are about noise and its various sources in CCD imaging but hold equally well for other imaging modalities.

4.4.1 Photon Noise

When the physical signal that we observe is based on light, then the quantum nature of light plays a significant role. A single photon at $\lambda = 500$ nm carries an energy of $E = h\nu = hc/\lambda = 3.97 \times 10^{-19}$ J. The noise problem arises from the fundamentally statistical nature of photon production. We cannot assume that in a given pixel for two consecutive but independent observation intervals of length T, the same number of photons will be counted. The probability distribution for p photons in an observation window of length T seconds is known to be of the Poisson type:

$$P(p \mid \sigma, T) = \frac{(\sigma T)^p e^{-pT}}{p!}, \tag{4.2}$$

where σ is the intensity parameter measured in photons per second. It is critical to understand that even if there were no other noise sources in the imaging chain, the statistical fluctuations associated with photon counting over a finite time interval T would still lead to a finite signal-to-noise ratio (SNR). For very bright signals, where σT exceeds 10^5, the noise fluctuations due to photon statistics can be ignored if the sensor has a sufficiently high saturation level.

4.4.2 Thermal Noise

An additional stochastic source of electrons in a CCD camera is thermal energy. Electrons can become increasingly agitated and consequently freed from the CCD material itself through thermal vibration and then, if trapped in the CCD well, can be indistinguishable from "true" photoelectrons. By cooling the CCD chip, it is possible to significantly reduce the number of thermal electrons that give rise to thermal noise, or *dark current*. The probability distribution of thermal electrons is also a Poisson process, where the rate parameter, τ, is an increasing function of temperature:

$$P(p \mid \tau, T) = \frac{(\tau T)^p e^{-pT}}{p!}. \tag{4.3}$$

4.4.3 On-Chip Electronic Noise

This noise originates in the process of reading the signal from the sensor, in this case through the field effect transistor (FET) of a CCD chip. The general form of the power spectral density of readout noise is

$$S(f) \propto \begin{cases} (2\pi f)^{-\beta} & f < f_{\min} & \beta > 0 \\ k & f_{\min} < f < f_{\max} \\ (2\pi f)^{-\alpha} & f > f_{\max} & \alpha > 0 \end{cases}, \qquad (4.4)$$

where α, β, and k are constants and f is the frequency at which the signal is transferred from the CCD chip to the subsequent circuits.

4.4.4 KTC Noise

Noise associated with the gate capacitor of an FET is termed *KTC noise* and can be nonnegligible. The output RMS value of this noise voltage is given by

$$n_{KTC} = \sqrt{\frac{kT}{C}}, \qquad (4.5)$$

where C is the FET gate switch capacitance, k is Boltzmann's constant, and T is the absolute temperature of the CCD chip measured in K. Proper electronic design that makes use, for example, of correlated double sampling and dual-slope integration can almost completely eliminate KTC noise.

4.4.5 Amplifier Noise

The standard model for this type of noise is additive, Gaussian, and independent of the signal. In modern well-designed electronics, amplifier noise is generally negligible. The most common exception to this is in color cameras where more amplification is used in the blue color channel than in the green channel or red channel, leading to more noise in the blue channel.

4.4.6 Quantization Noise

Quantization noise is inherent in the amplitude quantization process and occurs in the analog-to-digital converter, ADC. The noise is additive and independent of the signal when the number of grayscale levels is usually more than the 16 levels corresponding to a 4-bit representation.

Quantization noise can usually be ignored as the total SNR of a complete system is typically dominated by the smallest SNR. In CCD cameras, this is photon noise.

4.5 DISTORTIONS IN IMAGES

A more pertinent problem in the photography scenario is that of distortion, commonly called *blur*. It is treated as a particular point spread function (PSF) that is convolved with the image. As the name suggests, the PSF is a type of a kernel that arises from one point and spreads out in a fading manner. The blurring of images is basically a continuous process; the PSF can also be presented in a continuous manner. The PSF can be represented as $h(m,n,\psi)$, where ψ represents the severity function of blur PSF. A PSF cannot take any arbitrary value. However, due to the physics of the underlying imaging system, both $f(m,n)$ and $g(m,n)$ cannot be negative. As a consequence, the PSFs need to be nonnegative and real-valued.

In addition to this, the imperfections in an image formation system normally act as passive operations on the data; that is, they do not absorb or generate energy. Consequently, all energy arising from a specific point in the original image should be preserved, yielding

$$\int_{-\infty}^{\infty} \int_{-\infty}^{\infty} h(m,n;\psi)\, dm\, dn = 1, \tag{4.6}$$

or in discrete form,

$$\sum_{m,n} h(m,n) = 1, \tag{4.7}$$

where m and n are the indices of an image pixel. Four commonly used blur PSFs are presented next.

4.5.1 Linear Motion Blur

Many types of motion blurs can be distinguished, all of which are due to relative motion between the recording device and the object. This can be in the form of translation, a rotation, a sudden change of scale, or some combination of these. Translation, which is the most significant source of motion blur, can be described as follows.

When the object translates at a constant velocity V at an angle of ϕ radians with the horizontal axis during the exposure interval $[0,T]$, the PSF is given by

$$h(m,n;L,\phi)=\begin{cases} \dfrac{1}{L}, & if\ \sqrt{m^2+n^2}\le \dfrac{L}{2}\ \ and\ \ \dfrac{m}{n}=-\tan\phi, \\[2mm] 0, & elsewhere \end{cases} \tag{4.8}$$

where $L = VT$ is the length of motion. The PSF is space invariant. However, if only part of the image is subjected to the translational motion, the overall distortion is spatially variant.

4.5.2 Uniform Out-of-Focus Blur

When a camera on a 2D imaging plane images a 3D object, some parts of the object are in focus whereas other parts are not. For a circular aperture-imaging device, the image of any portion of the object is in the form of a circular disk known as *circle of confusion* (CoC). The degree of defocus (diameter of the CoC) depends on the focal length F and the aperture number k of the lens, and the distance P between camera and object. If the camera is focused sharply at an object at distance S, the diameter of the CoC, denoted by $C(P)$, is given as

$$C(P)=\begin{cases} \dfrac{FS}{k(S-F)}-\dfrac{F^2S}{kP(S-F)}-\dfrac{F}{k} & for \quad S<P<\infty \\[3mm] \dfrac{-FS}{k(S-F)}+\dfrac{F^2S}{kP(S-F)}+\dfrac{F}{k} & for \quad F<P<S \end{cases} \tag{4.9}$$

The PSF of this uniform out-of-focus blur with a radius of $R = C(P)/2$ is then given by

$$h(m,n;R)=\begin{cases} \dfrac{1}{\pi R^2}, & if\ \sqrt{m^2+n^2}\le R \\[2mm] 0, & elsewhere \end{cases} \tag{4.10}$$

4.5.3 Atmospheric Turbulence Blur

Atmospheric turbulence is a severe limitation for remote sensing and aerial imaging systems. Though this blur is manifested due to several factors, for a long-term exposure case, the blur PSF can be approximated as a Gaussian function as follows

$$h(m,n;\sigma_G) = C\,exp\left\{-\frac{m^2+n^2}{2\sigma_G^2}\right\}, \qquad (4.11)$$

where σ_G represents the severity of the blur and C is a constant, selected in such a way that the PSF satisfies its necessary conditions.

4.5.4 Scatter Blur

This blur occurs for specialized systems in which the incident imaging quanta are reflected by the system structure or other incident quanta. This phenomenon is more prominent in the case of x-ray and ultrasonic imaging. For such cases, the PSF is defined as follows:

$$h(m,n;\beta_m) = \frac{C}{\left[\beta_m^2+(m^2+n^2)\right]^{\frac{3}{2}}}, \qquad (4.12)$$

where β_m determines the sensitivity of the blur, and is a function of the distance between the radiated object and the detector.

4.6 ALGORITHMIC ACCOUNT

All of the aforementioned anomalies in images are essentially mathematical operations that, in general, take place at pixel level. However, these may expand into a group of interconnected pixels, as in the case of the scaling operation or the blurring process, where a whole neighborhood is affected by the operation. The following section covers some of the algorithmic aspects of these operations.

4.6.1 Affine Operations

Any of the three affine operators mentioned earlier can be used as matrix operations very easily in MATLAB®. However, for any other programming environment, these would be achieved through the use of nested loops with equations as shown in Table 4.1. Translation is probably the easiest of all because it only involves a reassignment of coordinate values.

The rotation algorithm, unlike the translation algorithm, can produce coordinates (x_2,y_2) that are not integers. In order to generate the intensity of the pixels at each integer position, different heuristics (or *resampling* techniques) may be employed. For example, two common methods include the following:

- Allow the intensity level at each integer pixel position to assume the value of the nearest noninteger neighbor (x_2,y_2).

- Calculate the intensity level at each integer pixel position based on a weighted average of the n nearest noninteger values. The weighting is proportional to the distance or pixel overlap of the nearby projections.

The latter method produces better results, but increases the computation time of the algorithm.

Scaling is probably the most complicated and involved operation of the TRS operators. Scaling compresses or expands an image along the coordinate directions. As different techniques can be used to subsample and zoom, each is discussed in turn.

Figure 4.3 shows four possibilities of scaling.

Figure 4.3(a) illustrates the two methods of subsampling. In the first, one pixel value within a local neighborhood is chosen (perhaps randomly) to be representative of its surroundings. (This method is computationally simple, but can lead to poor results if the sampling neighborhoods are too large.) In the figure, in the neighborhood of 2 × 2,

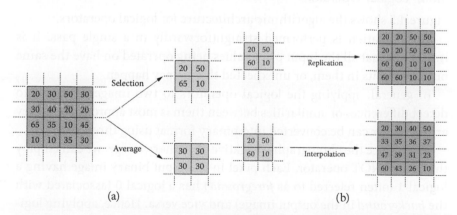

FIGURE 4.3 Scaling methods. (a) Reducing the size by half, (b) doubling the size.

the top-left pixel is selected for the new image pixel value. The second method interpolates between pixel values within a neighborhood by taking a statistical sample (such as the average) of the local intensity values.

An image (or regions of an image) can be zoomed either through pixel replication or interpolation. The first method in Figure 4.3(b) shows how pixel replication simply replaces each original image pixel by a group of pixels with the same value (where the group size is determined by the scaling factor; in this case, it is 2×2). Alternatively, interpolation of the values of neighboring pixels in the original image can be performed in order to replace each pixel with an expanded group of pixels. Most implementations offer the option of increasing the actual dimensions of the original image, or retaining them and simply zooming a portion of the image within the old image boundaries. In the figure, the interpolation values are calculated as approximate samples on a straight line between the two extreme points defined by the smaller image.

The neighborhood selection can also be straightforward or based on some predefined scheme, and can be algorithmically defined as $T_x[x(i:M, j:N,:)]$ and $T_y[x(i:M, j:N,:)]$, where T_x and T_y are the mathematical equivalents of the selector strategy for the source image pixels range $x(i:M, j:N,:)$. The resulting intensity value can be represented by T_m, which could be either "selection" or "averaging" in this case.

In general, the algorithmic logic should be somewhat like the one shown in Figure 4.4.

4.6.2 Logical Operators

Figure 4.5 shows the algorithmic architecture for logical operators.

The operation is performed straightforwardly in a single pass. It is important that all the input pixel values being operated on have the same number of bits in them, or unexpected things may happen.

In general, applying the logical operators to two images in order to detect differences or similarities between them is most appropriate if they are binary or can be converted into binary format using thresholding.

To produce the photographic negative of a binary image, we can employ the logical NOT operator. Each pixel in the input binary image having a logical 1 (often referred to as *foreground*) has a logical 0 (associated with the *background* in the output image) and vice versa. Hence, applying logical NOT to a binary image changes its polarity. The logical NOT can also be used for a grayscale image, in which case the resulting value for each

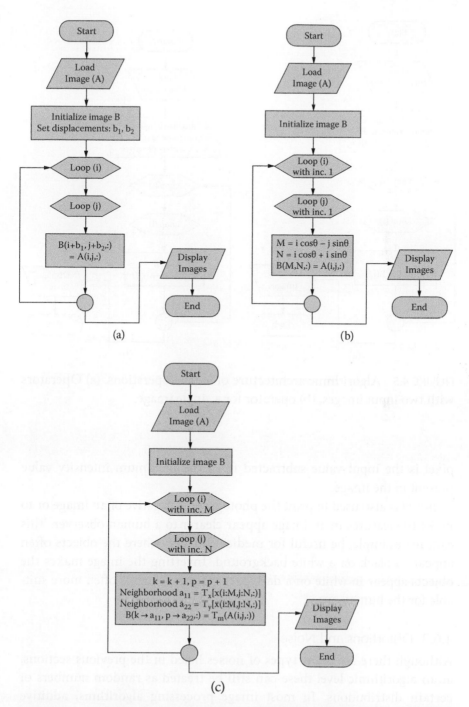

FIGURE 4.4 Algorithmic representations of the TRS operators. (a) Translation, (b) rotation, (c) scaling.

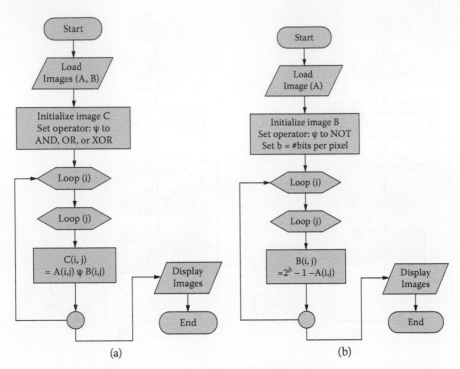

FIGURE 4.5 Algorithmic architecture of logical operations. (a) Operators with two input images, (b) operator for a single image.

pixel is the input value subtracted from the maximum intensity value present in the image.

Invert is also used to print the photographic negative of an image or to make the features in an image appear clearer to a human observer. This can, for example, be useful for medical images, where the objects often appear in black on a white background. Inverting the image makes the objects appear in white on a dark background, which is often more suitable for the human eye.

4.6.3 Distortions and Noise

Although there are many types of noises listed in the previous sections, at an algorithmic level these can still be treated as random numbers of certain distributions. In most image processing algorithms, additive Gaussian noise is the first choice. From Section 4.4, photon noise and amplifier noise are special cases of Gaussian noise, and the others can be

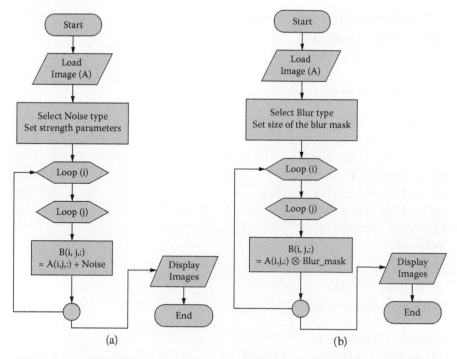

FIGURE 4.6 Algorithmic structures of noise and blur addition to images.

represented either by the Poisson distribution or spotted randomness. The Poisson distribution is used in a limited number of algorithms, especially for thermal imaging and atmospheric imaging for remote sensing applications. However, in the remaining types, a spotted noise is added and is usually called *salt-and-pepper* noise. The name is appropriate because it only changes the intensity at random places to either a very bright (salt) or a very dark (pepper) pixel.

Blur, on the other hand, is more involved than noise. It requires a specific PSF represented as a smaller matrix as the manifestation of the blurring function. This small matrix is then convolved with the source image to produce the blurred version. The mechanics of convolution will be discussed in Chapter 6. Figure 4.6 represents the algorithms for introducing noise and blur in an image.

4.7 MATLAB CODE

In this section, the MATLAB code used to generate all the images in this chapter is presented.

4.7.1 Affine and Logical Operators

```
clear all; close all
x=imread('lake.jpg');
imshow(x)
[r,c,s] = size(x) ;

% Image Translation
y1 = x(1:r1,1:c1,:) ;
y2 = x(1:r1,c1+1:c,:) ;
y3 = x(r1+1:r,1:c1,:) ;
y4 = x(r1+1:r,c1+1:c,:) ;
yy = [ y4 y3 ; y2 y1 ];
figure ; imshow(yy)
axis on

% Image Rotation
y=imrotate(x,45);
figure ; imshow(y)
z = imrotate(y,-45) ;
figure ; imshow(z)
[size(x) ; size(y) ; size(z) ]

% Image Scaling (resizing)
zz = imresize(x,[r/2 c/2]);
figure ; imshow(zz)
axis on
zz = imresize(x,[r/4 c/4]);
figure ; imshow(zz)
axis on
```

The preceding code generated the images in Figure 4.1. Similarly, the images in Figure 4.2 were generated by the following code:

```
clear all; close all
x=imread('lake.jpg');
imshow(x)
[r,c,s] = size(x) ;
[m,n] = ginput() ;
g(min(n):max(n),min(m):min(m),:) = 1 ;
g = g * 255 ;
figure ; imshow(g)
not_g = max(max(max(g)))-g;
```

```
% Image Inversion (NOTing)
not_x = 255-x;
figure ; imshow(not_x)

% Image AND/NAND-ing
g1 = double(double(x) .* g) ;
figure ; imshow(g1*0.00005)
mg = max(max(max(g1))) ;
figure ; imshow((mg-g1)*0.000008)

% Image OR/NOR-ing
g2 = double(double(x) + g) ;
figure ; imshow(g2*0.0005)
mg = max(max(max(g2))) ;
figure ; imshow((mg-g2)*0.003)

% Image XOR/XNOR-ing
g3 = double(x) .* not_g + double(not_x) .* double(g) ;
figure ; imshow(g3*0.00000005)
mg3=max(max(max(g3)));
figure ; imshow((mg3-g3)*0.00000005)
```

4.7.2 Noise in Images

The code for generating noisy images with Gaussian and salt-and-pepper noises with a strength parameter is as follows:

```
clear all; close all
x=imread('flowers001.jpg');
imshow(x)
[r,c,s] = size(x) ;

% Gaussian Noise
xn1 = imnoise(x,'gaussian',0,0.1);
figure ; surfl(xn1)
xn2 = imnoise(x,'gaussian',0,1);
figure ; imshow(xn2)

% Salt & Pepper Noise
xn3 = imnoise(x,'salt & pepper',0.05);
figure ; surfl(xn3)
xn4 = imnoise(x,'salt & pepper',0.3);
figure ; imshow(xn4)
```

(a)   (b)

(c)   (d)

FIGURE 4.7 Effect of added noise. (a) Gaussian noise with variance 0.1, (b) Gaussian noise with variance 1, (c) salt-and-pepper noise with parameter D = 0.05, (d) salt-and-pepper noise with parameter D = 0.3.

The resulting images are shown in Figure 4.7.

4.7.3 Blur in Images

The code for the blurring of images is as follows, and the resulting images are shown in Figure 4.8.

```
clear all; close all
x=imread('horses.jpg');
imshow(x)
[r,c,s] = size(x) ;

% Motion Blur
f1 = fspecial('motion',15,45) ;
figure ; surfl(f1)
xb1 = imfilter(x,f1);
figure ; imshow(xb1)
```

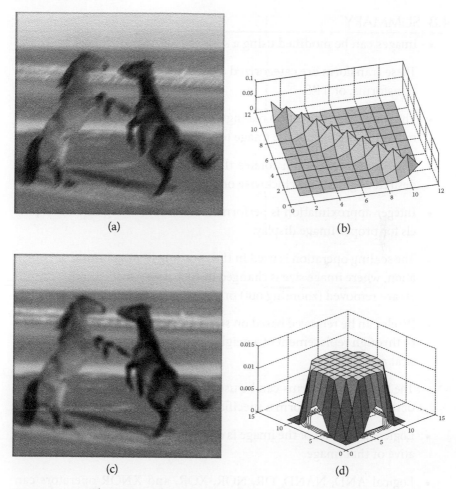

(a)

(b)

(c)

(d)

FIGURE 4.8 Blurring of images and the blurring masks. (a) Image blurred by "motion" blur representing motion in a 15-pixel neighborhood and at 45°, (b) the corresponding blurring function, (c) image blurred by "out-of-focus" blur with the circle of confusion (CoC) radius of 5 pixels, (d) corresponding blurring function.

```
% Out of Focus Blur
f2 = fspecial('disk',5) ;
figure ; surfl(f2)
xb2 = imfilter(x,f2);
figure ; imshow(xb2)
```

4.8 SUMMARY

- Images can be modified using a series of affine transforms.

- These transform are categorized as translation, rotation, and scaling transforms, or TRS.

- The translation operation reassigns the coordinate values, essentially moving some part of the image to a different place.

- The rotation operation rotates the image or part of the image by a desired angle, either clockwise or counterclockwise.

- Integer approximation is performed for the resulting fractional pixels for proper image display.

- The scaling operation is used in the zooming in or zooming out operation, where image size is changed in such a way that either some pixels are removed (zooming out) or new pixels are added (zooming in).

- Pixels can be removed based on some predefined selection procedure or through replacement of a neighborhood with an equivalent statistical measure.

- The logical transforms can be used to mask out a particular portion of the image to perform a specific operation.

- Logical inversion of the image is equivalent to the photographic negative of that image.

- Logical AND, NAND, OR, NOR, XOR, and XNOR operators can perform various localized selections or maskings on an image by suitably selected masking images.

- Noise in the image is the addition of random values to the image.

- Although physically there are many types of image noise, the algorithmic versions usually rely on Gaussian and salt-and-pepper noise to simulate natural phenomena.

- Blurring refers to other distortions in the image such as motion blur, out of focus, atmospheric effects, etc.

- Algorithmically speaking, blur is another 2D matrix (usually much smaller than the source image) that is then convolved with the image to produce the blurring effect.

4.9 EXERCISES

1. Image reflection is a special case of the translation operation. Write a program to read an image and produce another image that contains the source image and its reflection connected to it. This resulting image should have the same height as the original image but twice the width.

2. Each time an image of size M × N is rotated by an angle θ ($\theta < 90°$), the size of the image changes. Derive two equations relating the new height (P) and new width (Q) to the angle of rotation and the original dimensions.

3. Derive the linear interpolation equations (using simple line equations) to map a 2 × 2 image into an $n \times n$ image with the four corners of this new image as the four values in the 2 × 2 image segment. [Hint: The resulting system of equations is to be designed for $(n - 2) \times (n - 2)$ pixels to be injected. The 22 segments can be assumed to have pixels a_{11}, a_{12}, a_{21}, and a_{22}.]

4. Load the image pout.tif from MATLAB's image collection. Isolate the pocket of the shirt that the child is wearing in the image. Invert it as a photographic negative, and replace it in the original image. Other than the reading and displaying of the image, this can be done with one or two statements in MATLAB.

5. Out-of-focus blur is a special case of low-pass filtering and can be used to remove noise from the image at the cost of some blurring. Use a blurring function of a radius 2 and experiment with images in Figures 4.7 (a) and (d), and give a detailed account of your findings.

CHAPTER **5**

Image Transforms

5.1 INTRODUCTION

IMAGE TRANSFORMS REFER TO the mathematical mapping of image pixels from the spatial domain to some other domain where certain other characteristics of the image can be seen in a better manner. An analogy to this is the spectrum of a signal in the 1D case. In the time domain, we cannot readily see the frequency components, but mapping the signal to its spectrum reveals the details about these components (frequency and amplitude). The basic principle behind any transform is to decompose a signal (or image) into certain components using some type of magnifying-glass function called the *kernel* or *basis function*. As such, for images, this basic principle can be written as follows:

$$F(a,b) = \sum_m \sum_n f(m,n)\psi(m,n,a,b). \qquad (5.1)$$

Several types of such transforms are available for images, such as the following:

- Fourier transform
- Wavelet transform
- Hough transform
- Discrete cosine transform (discussed in Chapter 8)

However, in this chapter, the emphasis is on the first three transforms in this list. Other types of transforms will be discussed when their applications are encountered during the course of the discussion.

5.2 DISCRETE FOURIER TRANSFORM (DFT) IN 2D

The Fourier transform is an important image processing tool that is used to decompose an image into its sine and cosine components. The output of the transformation represents the image in the Fourier, or frequency, domain, and the input image is the spatial domain equivalent. In the Fourier domain image, each point represents a particular two-dimensional frequency component.

The DFT is the sampled Fourier transform and therefore does not contain all frequencies forming an image, but only a set of samples that is large enough to fully describe the spatial domain image. The number of frequencies corresponds to the number of pixels in the spatial domain image; that is, the images in the spatial and Fourier domains are of the same size. For an image of size $M \times N$, the two-dimensional DFT is given by

$$F(u,v) = \frac{1}{N^2} \sum_{m=0}^{M-1} \sum_{n=0}^{N-1} f(m,n) e^{-j2\pi\left(\frac{um}{M} + \frac{vn}{N}\right)}, \quad (5.2)$$

where $f(m,n)$ is the image in the spatial domain and the exponential term is the basis function corresponding to each point $F(u,v)$ in the Fourier space. The equation can be interpreted as follows: the value of each point $F(u,v)$ is obtained by multiplying the spatial image by the corresponding basis function and summing the result.

The basis functions are sine and cosine waves with increasing frequencies; that is, $F(0,0)$ represents the DC component of the image, which corresponds to the average brightness, and $F(N-1,N-1)$ represents the highest frequency.

An inverse transformation is also possible and is given by

$$f(m,n) = \frac{1}{N^2} \sum_{u=0}^{M-1} \sum_{v=0}^{N-1} F(u,v) e^{j2\pi\left(\frac{um}{M} + \frac{vn}{N}\right)}. \quad (5.3)$$

The ordinary one-dimensional DFT has MN complexity. This can be reduced to $M \log_2 N$ or $N \log_2 M$ (whichever is smaller) if we employ the *fast Fourier transform* (FFT) to compute the one-dimensional DFTs. This is a significant improvement, in particular, for large images. There are various forms of the FFT, and most of them restrict the size of the input image that may be transformed, often to $N = 2^n$, where n is an integer. The mathematical details are well described in the literature.

The Fourier transform produces a complex number-valued output image that can be displayed with two images, either with the *real* and *imaginary* part or with the *magnitude* and *phase*. In image processing, often only the magnitude of the Fourier transform is displayed, as it contains most of the information of the geometric structure of the spatial domain image. However, if we want to retransform the Fourier image into the correct spatial domain after some processing in the frequency domain, we must make sure to preserve both magnitude and phase of the Fourier image.

The Fourier transform is used if we want to access the geometric characteristics of a spatial domain image. Because the image in the Fourier domain is decomposed into its sinusoidal components, it is easy to examine or process certain frequencies of the image, thus influencing the geometric structure in the spatial domain.

In most implementations, the Fourier image is shifted in such a way that the DC value (i.e., the image mean), $F(0,0)$, is displayed in the center of the image. The further away from the center an image point is, the higher is its corresponding frequency. Figure 5.1 shows an example of application of the Fourier transform to images.

It can be seen that the R, G, and B layers are not providing more information than any single layer; therefore, it is always advisable for Fourier transform applications to work with grayscale images. Figure 5.2 shows the effect of changes in orientation of the original image.

5.3 WAVELET TRANSFORMS

Unlike the Fourier transform, in which the basis functions cover the entire signal range with frequency variation only, the wavelet transform performs better decomposition of a signal into its components because in this case, the basis functions vary in both frequency range (called *scale*) and spatial range. High-frequency basis covers a smaller area, whereas low-frequency basis covers a larger area. This transform is more appropriate for nonstationary signals and provides nonuniform partitions of frequency and spatial ranges.

In Fourier analysis, a signal is broken up into sine and cosine waves of different frequencies, and it effectively rewrites a signal in terms of different sine and cosine waves. Wavelet analysis does a similar thing: It takes a mother wavelet, and the signal is translated into shifted and scaled versions

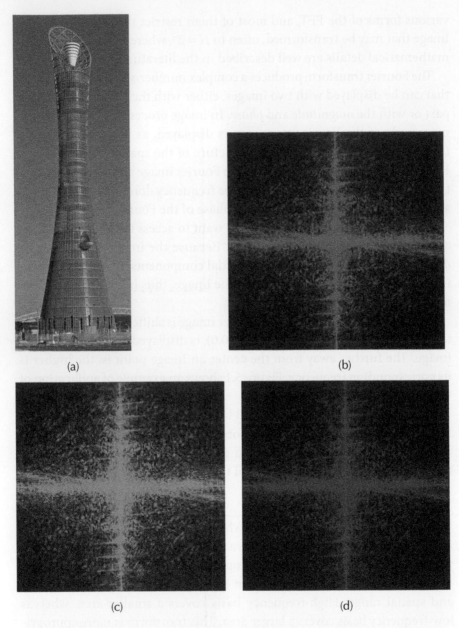

(a)

(b)

(c)

(d)

FIGURE 5.1 (See color insert following Page 204.) Application of Fourier transform to images. (a) Original RGB image, (b) R, (c) G, (d) B components of the Fourier-transformed image.

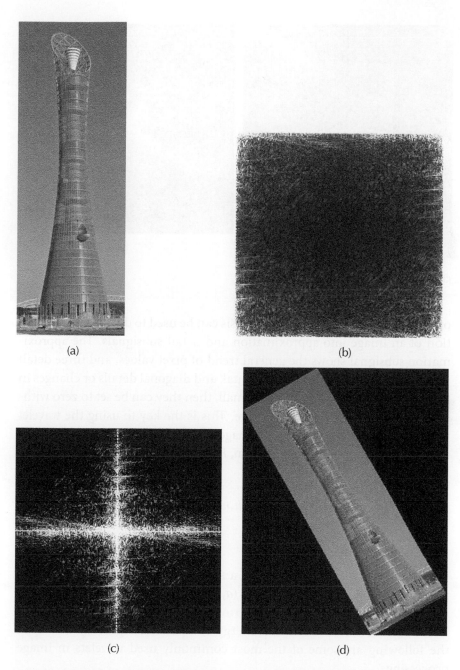

(a)

(b)

(c)

(d)

FIGURE 5.2 Effect of changes in orientation of the original image. (a) Original grayscale image, (b) FFT of (a) without shifting the components, (c) FFT of (a) after shifting, (d) rotated image by 30°, (e) FFT of (d) without shifting the components, (f) FFT of (d) after shifting.

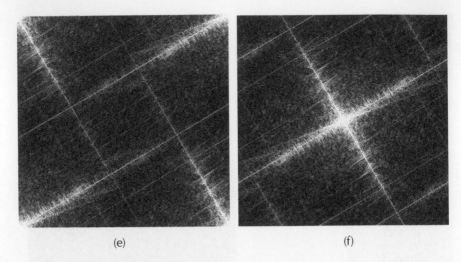

(e)   (f)

FIGURE 5.2   (Continued)

of this mother wavelet. Wavelet analysis can be used to divide the information of an image into approximation and detail subsignals. The approximation subsignal shows the general trend of pixel values, and three detail subsignals show the vertical, horizontal, and diagonal details or changes in the image. If these details are very small, then they can be set to zero without significantly changing the image. This is the key to using the wavelet transform in applications such as image filtering and image compression.

The 2D discrete wavelet transform, $F(u,v)$ of an image $f(m,n)$ is given by

$$F(a_1,b_1,a_2,b_2)=\sum_{m=0}^{M-1}\sum_{n=0}^{N-1} f(m,n)\frac{1}{\sqrt{b_1 b_2}}\psi\left(\frac{m-a_1}{b_1},\frac{n-a_2}{b_2}\right), \qquad (5.6)$$

where $\frac{1}{\sqrt{b_1 b_2}}\psi(\frac{m-a_1}{b_1},\frac{n-a_2}{b_2})$ represents a specific type of wavelet with scaling and shifting in x and y axes as (a_1,b_1) and (a_2,b_2), respectively. There have been many designs related to newer and better mother wavelets. These mother wavelets differ in energy compaction capability. However, the following are some of the most commonly used wavelets in image processing:

- "haar": Haar wavelet

- "db":   Daubechies wavelets

- "sym": Symlets

- "coif": Coiflets

- "bior": Biorthogonal wavelets

- "rbio": Reverse biorthogonal wavelets

- "meyr": Meyer wavelet

- "dmey": Discrete Meyer wavelet

- "gaus": Gaussian wavelets

- "mexh": Mexican hat wavelet

- "morl": Morlet wavelet

- "cgau": Complex Gaussian wavelets

- "cmor": Complex Morlet wavelets

The inverse wavelet transform is used to reconstruct the function from its wavelet representation. This is represented as follows:

$$f(m,n) = \sum_{a_1=0}^{A_1-1} \sum_{b_1=0}^{B_1-1} \sum_{a_2=0}^{A_2-1} \sum_{b_2=0}^{B_2-1} F(a_1,b_1,a_2,b_2) \frac{1}{\sqrt{b_1 b_2}} \psi^{-1}\left(\frac{m-a_1}{b_1}, \frac{n-a_2}{b_2}\right). \quad (5.7)$$

Discrete wavelet analysis is usually computed using the concept of filter banks. Filters of different cutoff frequencies analyze the signal at different scales. Resolution is changed by the filtering, the scale is changed by upsampling and downsampling. If a signal is put through two filters:

1. High-pass filter—High-frequency information is retained, low-frequency information is lost.

2. Low-pass filter—Low-frequency information is retained, high-frequency information is lost.

Then the signal is effectively decomposed into two parts, a detailed part (high-frequency), and an approximation part (low-frequency). The subsignal produced from the low filter will have a highest frequency equal to half that of the original. According to Nyquist sampling, this change in frequency range means that only half of the original samples need to

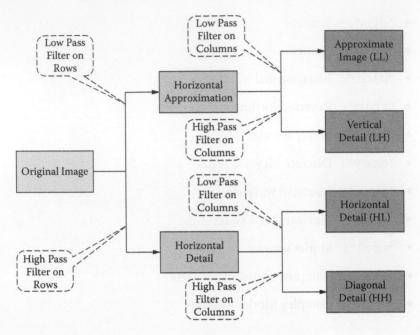

FIGURE 5.3   Basic principle of wavelet transformation.

be kept in order to perfectly reconstruct the signal. More specifically, this means that upsampling can be used to remove every second sample. The scale has now been doubled. The resolution has also been changed; the filtering made the frequency resolution better, but reduced the time resolution.

The approximation subsignal can then be put through a filter bank, and this is repeated until the required level of decomposition has been reached. These ideas are shown in Figure 5.3.

At every level, four subimages are obtained: the approximation, the vertical detail, the horizontal detail, and the diagonal detail. Wavelet analysis has revealed how the image changes vertically, horizontally, and diagonally. This process can be further repeated for the approximate subsignal for many levels until a certain set of criteria is met, which depends on the application. With each decomposition, a new approximation is obtained that is much smaller in size. The decomposition pattern for images is shown in Figure 5.4.

Figure 5.5 shows a typical example of application of decomposition with the wavelet transform. The mother wavelet used in this example is the "Daubechies" type.

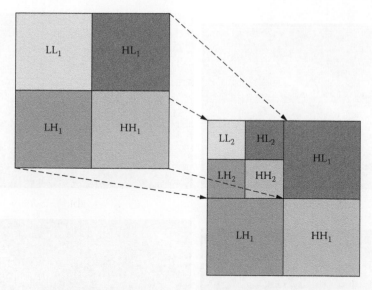

FIGURE 5.4 Multilevel decomposition with wavelet transforms.

5.4 HOUGH TRANSFORM

The conventional Hough transform can be commonly used for the detection of regular curves such as lines, circles, ellipses, etc. Despite its domain restrictions, the conventional Hough transform has many applications. Its main advantage is that it is tolerant of gaps in feature boundary descriptions and is relatively unaffected by image noise. The fundamental idea of the Hough transform is based on the polar coordinate representation of the line. If a line exists in the image, it will be in accordance with a polar equation given by

$$\rho = m\cos\theta + n\sin\theta, \tag{5.8}$$

where ρ is the length of a normal vector from the origin to this line and θ is the orientation of ρ with respect to the x-axis. For any point (m,n) on this line, r and θ are constant. Hence, for every (m,n) coordinate pair on the line, there exists a value of θ and ρ that is an entirely different manifestation of the image. Many people when encountering the Hough transform think that it is a fancy edge detector, but it goes beyond that by providing the connectivity information of the lines as well. For instance, in the case of a rectangular shape, there are four lines in which case the lines of one pair start at the common point and then meet the corresponding lines

(a) (b)

(c) (d)

FIGURE 5.5 Wavelet decomposition of the football image. (a) Original RGB image (x); (b) first decomposition into [xa, xh, xv, xd]; (c) second decomposition into [xaa, xhh, xvv, xdd]; (d) overall decomposition.

from the other pair, after which each line returns to its own paired line that it started from in the first place. The shape-related features are quite independent of noise and distortion present in the image and, therefore, the technique is very robust. This is shown in Figure 5.6.

In an image analysis context, the coordinates of the points of edge segments [i.e., (m_i, n_i)] in the image are known and therefore serve as constants in the parametric line equation, and ρ and θ are the unknown variables we seek. If we plot the possible (ρ, θ) values defined by each (m_i, n_i) point, the resulting points in Cartesian image space map to curves (i.e., sinusoids) in the polar Hough parameter space. This point-to-curve transformation

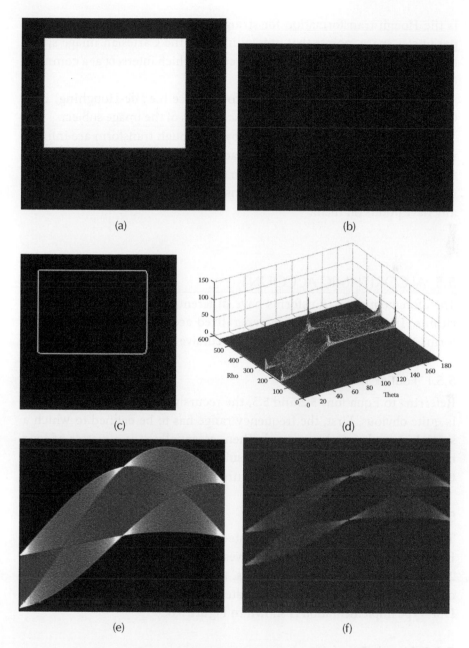

FIGURE 5.6 The Hough transform. (a) Original RGB image, (b) inverse grayscale image, (c) edge detection using Canny operator, (d) mesh plot for the Hough transform, (e) image from display of the Hough transform, (f) contour plot of the Hough transform.

is the Hough transformation for straight lines. When viewed in Hough parameter space, points that are collinear in the Cartesian image space become readily apparent as they yield curves which intersect at a common (ρ,θ) point.

Mapping back from Hough transform space (i.e., de-Houghing) into Cartesian space yields a set of line descriptions of the image subject.

Note also that the lines generated by the Hough transform are infinite in length. If we wish to identify the actual line segments that generated the transform parameters, further analysis of the transformed image is required in order to see which portions of these infinitely long lines actually have points on them.

5.5 ALGORITHMIC ACCOUNT

The aforementioned transforms have different implementations as algorithms. The following is a very generalized account of how these transforms can be implemented as general recursive procedures.

5.5.1 Fourier Transform

Referring to Equations 5.2 and 5.3, the recursive nature of the algorithm is quite obvious. First, the frequency range has to be defined to which a spatial image has to be mapped. Then, essentially, it is a matter of four nested loops to produce the final result. The same is true for the inverse transformation. This is shown in Figure 5.7.

5.5.2 Wavelet Transform

The wavelet transform is implemented quite differently as filter banks. Because breaking the image into halves is equivalent to downsampling by 2, first in rows and then in columns, the main implementation requires only a low-pass filter, a high-pass filter, and a downsampler. The actual DSP-type implementation is shown in Figure 5.8.

5.5.3 Hough Transform

It is easier to implement the Hough transform with three nested loops and for known edges. Therefore, it is usually advisable to do a simple Sobel- or Canny-type edge detection based on thresholding before applying the transform. These operators will be discussed in Chapter 9. Figure 5.9

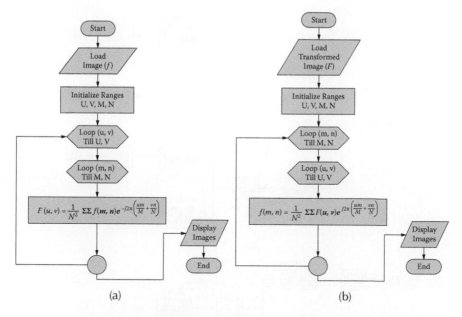

FIGURE 5.7 Fourier transform algorithmic implementation. (a) Simple transform, (b) inverse transform.

shows a typical implementation logic of the Hough transform. The loop for θ is arbitrary and can be initiated to cover a range of angles in the line of sight of a ρ line.

5.6 MATLAB® CODE

The MATLAB code used to produce the images in Figures 5.1, 5.2, 5.5, and 5.6 is presented in this section.

5.6.1 Fourier Transform

The first piece of code presented is related to Figure 5.1. Note the trick utilized in displaying the colors. A word of caution in this respect when using the function fftshift(). If the zero setting of individual layers is done before this function is used, the layers are also shifted and the colors will appear in single-layer shifted manner (i.e., R will become B, G will become R, and B will become G).

```
close all
clear all
x = imread('tower.jpg') ;
figure ; imshow(x) ;
```

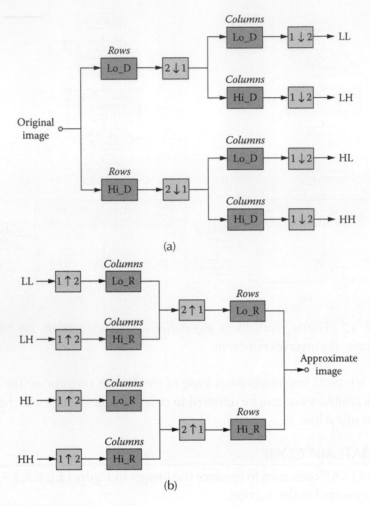

FIGURE 5.8 Filter bank implementation of the wavelet transform. (a) Wavelet decomposition, (b) wavelet composition. Here Lo_D is a low-pass filter for decomposition, Lo_R is a low-pass filter for reconstruction, Hi_D is a high-pass filter for decomposition, Hi_R is a high-pass filter for reconstruction, and blocks with arrows represent down or upsampling.

```
% RGB FFT (Figure 5.1)
X = fft2(x,1024,1024) ;
Y = fftshift(abs(X)) ;
YR = Y ; YR(:,:,[2 3]) = 0 ;
YG = Y ; YG(:,:,[1 3]) = 0 ;
YB = Y ; YB(:,:,[1 2]) = 0 ;
figure ; imshow(YR*0.0001)
```

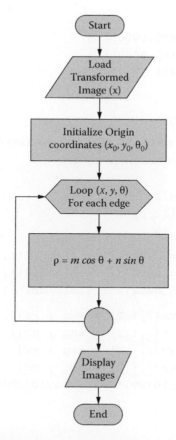

FIGURE 5.9   Algorithm for Hough transform.

```
figure ; imshow(YG*0.0001)
figure ; imshow(YB*0.0001)

% Grayscale FFT (Figure 5.2)
xg = rgb2gray(x) ;
figure ; imshow(xg) ;
XG = fft2(xg,1024,1024) ;
XGM=(abs(XG)) ;
figure ; imshow(XGM*0.00005) ;
XGMS = fftshift(XGM) ;
figure ; imshow(XGMS*0.00005) ;

%Effect of Rotation (Figure 5.2)
xr = imrotate(xg,30) ; %rotation by 30 degrees
figure ; imshow(xr) ;
XR = fft2(xr,1024,1024) ;
```

```
XRM=(abs(XR)) ;
figure ; imshow(XRM*0.00005) ;
XRMS = fftshift(XRM*0.00005) ;
figure ; imshow(XRMS) ;
```

5.6.2 Wavelet Transform

Figure 5.5 shows two levels of wavelet decomposition of the football image. The following code generated these images:

```
clear all ; close all
x = imread('football.jpg') ;
figure ; imshow(x) ;

[xar,xhr,xvr,xdr] = dwt2(x(:,:,1),'db2');
[xag,xhg,xvg,xdg] = dwt2(x(:,:,2),'db2');
[xab,xhb,xvb,xdb] = dwt2(x(:,:,3),'db2');

xa(:,:,1) = xar ; xa(:,:,2) = xag ; xa(:,:,3) = xab ;
xh(:,:,1) = xhr ; xh(:,:,2) = xhg ; xh(:,:,3) = xhb ;
xv(:,:,1) = xvr ; xv(:,:,2) = xvg ; xv(:,:,3) = xvb ;
xd(:,:,1) = xdr ; xd(:,:,2) = xdg ; xd(:,:,3) = xdb ;
X1 = [ xa*0.003 log10(xv)*0.3 ; log10(xh)*0.3
log10(xd)*0.3 ] ;
figure ; imshow(X1)

[xaar,xhhr,xvvr,xddr] = dwt2(xa(:,:,1),'db2');
[xaag,xhhg,xvvg,xddg] = dwt2(xa(:,:,2),'db2');
[xaab,xhhb,xvvb,xddb] = dwt2(xa(:,:,3),'db2');
xaa(:,:,1) = xaar ; xaa(:,:,2) = xaag ; xaa(:,:,3) = xaab ;
xhh(:,:,1) = xhhr ; xhh(:,:,2) = xhhg ; xhh(:,:,3) = xhhb ;
xvv(:,:,1) = xvvr ; xvv(:,:,2) = xvvg ; xvv(:,:,3) = xvvb ;
xdd(:,:,1) = xddr ; xdd(:,:,2) = xddg ; xdd(:,:,3) = xddb ;

X11 = [ xaa*0.001 log10(xvv)*0.3 ; log10(xhh)*0.3
log10(xdd)*0.3 ] ;
figure ; imshow(X11)

[r,c,s] = size(xv) ;
figure ; imshow([X11(1:r,1:c,:) xv*0.05 ; xh*0.05
xd*0.05 ])
```

Note the methodology in displaying the RGB images, although the processing is done on each layer separately.

5.6.3 Hough Transform

The following code generated the images in Figure 5.6:

```
clear all
close all
x = imread('mask1.jpg') ;
x = x(201:400,401:600);
figure ; imshow(x)
y = 255-x ;
figure ; imshow(y)
z = edge(y,'canny',0.7);
figure ; imshow(z)
[h,r,t] = hough(z) ;
figure ; mesh(h)
axis on
xlabel('theta')
ylabel('rho')
figure
h1 = h(end:-1:1,:) ;
imshow(h1(51:420,:)*0.1)
axis on
```

5.7 SUMMARY

- Image transforms decompose an image into a data space where characteristics that are otherwise not visible in the original image can be manifested.

- Usually, the transformation is bidirectional, and the original image can be restored using the inverse transformation.

- The Fourier transform is a very well-established technique in mathematics and is useful in describing a signal in terms of its frequency components.

- In an image, the frequency components are related to both the height and width of the image.

- Any sudden change in the grayscale values represents high-frequency components.

- Consistent gradients or constant grayscale values are related to low-frequency components.

- Computationally, a variant of the discrete Fourier transform is utilized, and it is called the *fast Fourier transform*, or FFT.

- The wavelet transform's function is similar to that of the Fourier transform, but it changes the view window from a large window to a smaller one; this is called *scaling*.

- The scaling decomposition helps to better isolate frequency components compared to the Fourier transform.

- Unlike simple sine/cosine basis functions of the Fourier transform, the wavelet transforms use specialized basis functions called *mother wavelets*.

- There are a number of these mother wavelets in use for various applications.

- Wavelet transforms are primarily used for noise removal and compression applications.

- The Hough transform is a classical technique, and is computationally expensive.

- It can provide information related to the connectivity of edges, which can be used for shape identification.

5.8 EXERCISES

1. Use the noisy image of Figure 4.7(b), and find its Fourier transform. Compare this transform with the transform for the original image. Answer the following questions based on your comparison:

   a. Which part of the spectrum is affected by the noise?

   b. What part of the image do you think has contributed to the central part of the transform?

   c. Can we remove some part of the spectrum to remove the noise?

2. Use the clown.mat file from MATLAB's image collection with the "Daubechies" wavelet. Perform wavelet decomposition iteratively for seven iterations. At what point did you notice that the approximation (LL) is no longer similar to the original image? In your opinion, what could be the reason for that?

3. In Figure 5.6, a rectangle is used whose shape is described using the Hough transform. Repeat this example with a circle [Figure EX5.1(a)] in the image instead of a rectangle.

   a. How many edges can be detected by the Hough transform?

   b. Develop a descriptive rule that can be used to define the shape, circle.

4. Repeat Question 3 for EX5.1(b).

5. Apply the Fourier transform on the blurred image of Figure 4.8(c), and compare it with the transform of the original horses.jpg image. Why is there a difference in the number of high-frequency components?

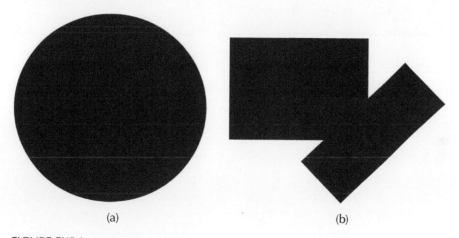

(a)   (b)

FIGURE EX5.1

CHAPTER 6

Spatial and Frequency Domain Filter Design

6.1 INTRODUCTION

IN IMAGE PROCESSING, FILTERS are mainly used to suppress either the high frequencies in the image, that is, smoothing the image, or the low frequencies, that is, enhancing or detecting edges in the image. An image can be filtered either in the frequency or in the spatial domain. In both cases, the intent is to replace some of the pixels with approximate pixels that can improve the overall visualization of the image. In the case of spatial domain filtering, the filtering operation is performed directly on the image without any transformation. Alternatively, in the frequency domain, the image is first transformed into the frequency domain using FFT-type algorithms and then appropriate filtering is applied. The spatial domain filters will usually rely on operations such as convolution, pixel-by-pixel multiplication, and summations, whereas the frequency domain processing requires masking, elimination, and prefiltering equalization type of operations.

6.2 SPATIAL DOMAIN FILTER DESIGN

The spatial filter design involves convolving the input image $f(m,n)$ with the filter function $h(m,n)$. This can be written as

$$g(m,n) = f(m,n) \otimes h(m,n). \tag{6.1}$$

This mathematical operation is identical to multiplication in the frequency space, but the results of the digital implementations vary, because we have to approximate the filter function with a discrete and finite kernel.

6.2.1 Convolution Operation

Discrete convolution is composed of three operations: *shift, multiply,* and *summation*, where we shift the kernel over the image, multiply its value with the corresponding pixel values of the image, and take the sum of these multiplications, obtaining the new pixel value. For a square kernel of size $M \times M$, we can calculate the output image with the following formula:

$$g(m,n) = \sum_{x=-\frac{M}{2}}^{\frac{M}{2}} \sum_{y=-\frac{M}{2}}^{\frac{M}{2}} h(x,y) f(m-x, n-y). \tag{6.2}$$

Various standard kernels exist for specific applications; the size and the form of the kernel determine the characteristics of the operation. Some of these kernels are discussed in this chapter.

6.2.2 Averaging/Mean Filter

Averaging or mean filtering is a simple, intuitive, and easy-to-implement method of *smoothing* images, that is, reducing the amount of intensity variation between one pixel and the next. It is often used to reduce noise in images because noise produces abrupt changes in pixel values.

The idea of mean filtering is simply to replace each pixel value in an image with the mean (average) value of its neighbors, including itself. This has the effect of eliminating pixel values that are unrepresentative of their surroundings. Often, a 3×3 square kernel is used, as shown in Figure 6.1(a), although larger kernels (e.g., 5×5 squares) can be used for greater smoothing.

Figure 6.2 show an example of using the mean filter for various types of noisy images. As can be seen from the images, the effect of randomness is reduced at the cost of "smoothing"; that is, the image becomes blurred after the filtering. Because the averaging or mean filter performs the averaging operation, it essentially smooths out the variations in the neighboring pixels. The removal of abrupt changes is equivalent to the removal of high-frequency components, and therefore, the averaging filter can also be thought of as a low-pass filter.

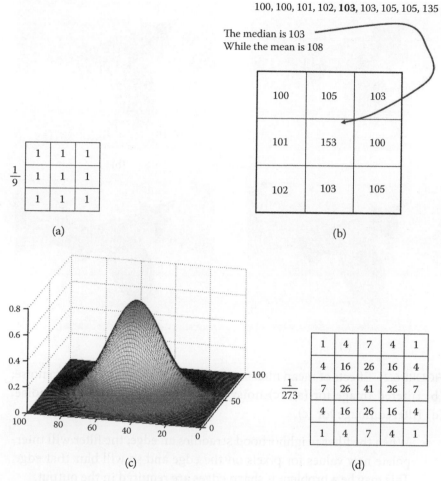

FIGURE 6.1 Various types of filter kernels. (a) 3 × 3 averaging filter kernel, (b) median kernel and result calculation, (c) 2D Gaussian distribution with mean (50%, 50%) and $s = 1$, (d) the matrix version of (c).

6.2.3 Median Filter

The median filter is normally used to reduce noise in an image, somewhat like the mean filter. However, it often does a better job than the mean filter of preserving useful detail in the image. The examples illustrated in Figure 6.2 show two main problems with mean filtering:

- A single pixel with a very unrepresentative value can significantly affect the mean value of all the pixels in its neighborhood.

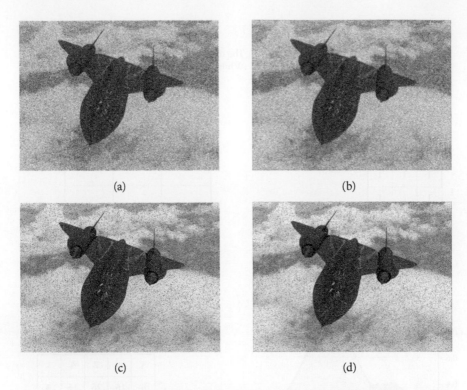

(a) (b)

(c) (d)

FIGURE 6.2 Effects of mean filtering. (a) Noisy image with Gaussian noise; (b) filtered image for (a); (c) noisy image with salt-and-pepper noise; (d) filtered image for (c).

- When the filter neighborhood straddles an edge, the filter will interpolate new values for pixels on the edge and so will blur that edge. This may be a problem if sharp edges are required in the output.

Both of these problems are resolved by the median filter, which is often a better filter for reducing noise than the mean filter, but it takes longer to compute. However, at the same time, it is more

The median filter considers each pixel in a neighborhood of a center pixel and replaces it with the *median* of the values in this neighborhood. The median is calculated by first sorting all the pixel values from the surrounding neighborhood in numerical order and then replacing the pixel being considered with the middle pixel value. (If the neighborhood under consideration contains an even number of pixels, the average of the two middle pixel values is used.) Figure 6.1(b) illustrates an example of the median filter. It can be seen with a little calculation that the median filter is very good in terms of removing the outliers in a neighborhood, something very desirable for removing salt-and-pepper noise from images. In the figure, the outlier value, 153, is replaced by the median of the neighborhood, 103. This is closer to the values of the neighborhood than the mean for the same neighborhood, 108.

By calculating the median value of a neighborhood rather than the mean, the median filter has two main advantages over the mean filter:

- The median is a more robust average than the mean, and so a single very unrepresentative pixel in a neighborhood will not affect the median value significantly.

- Because the median value must actually be the value of one of the pixels in the neighborhood, the median filter does not create new unrealistic pixel values when the filter straddles an edge. For this reason, the median filter is much better at preserving sharp edges than the mean filter.

Figure 6.3 shows the usage of the median filter.

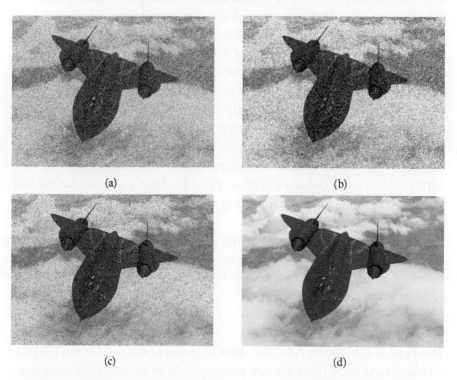

(a)

(b)

(c)

(d)

FIGURE 6.3 Effects of median filtering. (a) Noisy image with Gaussian noise; (b) filtered image for (a); (c) noisy image with salt-and-pepper noise; (d) filtered image for (c).

One of the major problems with the median filter is that it is relatively expensive and complex to compute. To find the median, it is necessary to sort all the values in the neighborhood in numerical order, and this is relatively slow, even with fast sorting algorithms such as *quicksort*.

6.2.4 Gaussian Smoothing

The Gaussian smoothing operator is a 2D convolution operator that is used to "blur" images and remove detail and noise. In this sense, it is similar to the mean filter, but it uses a different kernel that represents the shape of a Gaussian ("bell-shaped") hump. The Gaussian distribution in 2D has the form:

$$h(m,n) = \frac{1}{2\pi\sigma^2} e^{-\frac{\left\{(m-m_0)^2+(n-n_0)^2\right\}}{2\sigma^2}}. \tag{6.3}$$

This distribution is shown in Figure 6.1(c). The idea of Gaussian smoothing is to use this 2D distribution as a point spread function, and this is achieved by convolving the kernel with the image under process. Because the image is stored as a collection of discrete pixels, we need to produce a discrete approximation to the Gaussian function before we can perform the convolution. In theory, the Gaussian distribution is nonzero everywhere, which would require an infinitely large convolution kernel, but in practice, it is effectively zero more than about three standard deviations from the mean, and so we can truncate the kernel at this point. Figure 6.1(d) shows a suitable integer-valued convolution kernel that approximates a 2D Gaussian with a σ of 1.0. Once a suitable kernel has been calculated, Gaussian smoothing can be performed using standard convolution methods.

The effect of Gaussian smoothing is to blur an image, in a similar fashion to the mean filter. The degree of smoothing is determined by the standard deviation of the Gaussian. Figure 6.4 shows an example of the application of Gaussian filtering.

6.2.5 Conservative Smoothing

This type of smoothing is conservative in terms of not modifying the pixels blindly. As such, there are two conditions to be tested before a replacement is made. First, a neighborhood is selected around the pixel under study. Typically, an eight-neighbor region is selected (i.e., all eight pixels around the underlying pixel). The minimum and maximum values

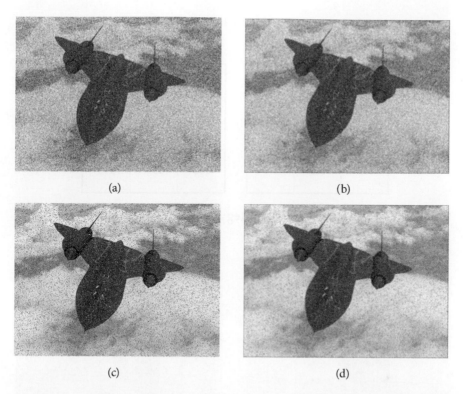

FIGURE 6.4 Effects of Gaussian filtering. (a) Noisy image with Gaussian noise; (b) filtered image for (a); (c) noisy image with salt-and-pepper noise; (d) filtered image for (c).

from this neighborhood are found, and the center pixel value is compared against these limiting values. If the value of the center pixel is above the maximum, it is set to the maximum value; if it is less than the minimum, it is set to the minimum value, else it is left as it is. Due to this selective replacement strategy, the chances of removing the outlier values, and consequently the speckle and salt-and-pepper noise, can be eliminated. Figure 6.5 shows this design strategy.

Figure 6.6 shows an example of the application of this filter. The effects can be fine-tuned by choosing a bigger neighborhood and experimenting with various signal-to-noise ratios as well as with different types of noise.

6.3 FREQUENCY-BASED FILTER DESIGN

Frequency filters process an image in the frequency domain. The image is Fourier-transformed, multiplied by the filter function, and then retransformed into the spatial domain. Attenuating high frequencies results in a

The maximum is 105, The minimum is 100

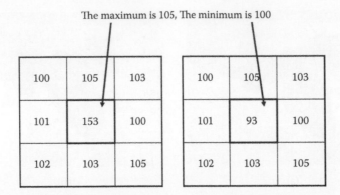

100	105	103
101	153	100
102	103	105

100	105	103
101	93	100
102	103	105

FIGURE 6.5   Conservative filter strategy.

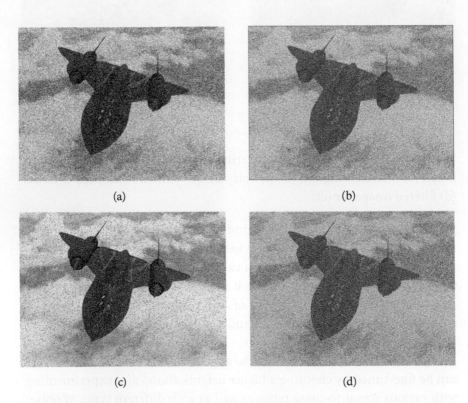

(a)

(b)

(c)

(d)

FIGURE 6.6   Example of conservative filtering. (a) Noisy image with Gaussian noise, (b) filtered image, (c) noisy image with salt-and-pepper noise, (d) filtered image.

smoother image in the spatial domain, whereas attenuating low frequencies enhance the edges. This is based on the property of the Fourier transform whereby convolution in the spatial domain becomes multiplication in the frequency domain, thus reducing the computation required for the filtering process by two thirds. The transformation and inverse transformations can be done as independent processes at the two ends of the process block. The general idea is as follows:

$$G(u,v) = F(u,v)H(u,v),$$

$$g(m,n) \Leftrightarrow G(u,v), \tag{6.4}$$

where $g(m,n)$ is the resulting image, $f(m,n)$ is the original image, and $h(m,n)$ is the filter kernel in the spatial domain. Correspondingly, $G(u,v)$, $F(u,v)$, and $H(u,v)$ are the frequency domain representations of the same. All frequency filters can also be implemented in the spatial domain and, if there exists a simple kernel for the desired filter effect, it is computationally less expensive to perform the filtering in the spatial domain. Frequency filtering is more appropriate if no straightforward kernel can be found in the spatial domain, and may also be more efficient.

In contrast to the frequency domain, it is possible to implement nonlinear filters in the spatial domain. In this case, the summations in the convolution function are replaced with some kind of nonlinear operator:

$$G(u,v) = O_{u,v}[F(u,v)H(u,v)], \tag{6.5}$$

where $O_{u,v}$ represents a nonlinear operation on the frequency space. The nonlinear function could be a specific shape filter, say Gaussian, or it may be some type of masking operation, or some algorithmic steps for iterative selection of values in an image.

The form of the filter function determines the effects of the operator. There are basically three different kinds of filters: *low-pass*, *high-pass*, and *bandpass* filters. A low-pass filter attenuates high frequencies and leaves low frequencies unchanged. The result in the spatial domain is equivalent to that of a smoothing filter as the blocked high frequencies correspond to sharp intensity changes, that is, to the fine-scale details and noise in the spatial domain image. A high-pass filter, on the other hand, yields edge enhancement or edge detection in the spatial domain, because edges contain many high frequencies. Areas of rather constant gray level consist of mainly low frequencies and are therefore suppressed. A bandpass filter attenuates very low and very high frequencies, but retains a middle range band of frequencies. Bandpass filtering can be used to enhance edges (suppressing low frequencies) while reducing the noise at the same time (attenuating high frequencies).

The most simple low-pass filter is the *ideal low-pass filter*. It suppresses all frequencies higher than the *cutoff frequency* f_0 and leaves smaller frequencies unchanged. Essentially, it is like a masking operation where a circle of all 1s is ANDed with the Fourier-transformed image. However, such a sharp cutoff causes the ringing effect when the image is reverse-transformed into the spatial domain. A better approach is to use a gradual cutoff shape such as a Gaussian mask. Figure 6.7 shows these filter masks and their effects.

Figure 6.8 is a repetition of Figure 6.7 for a salt-and-pepper noise-corrupted image.

The same principles apply to high-pass filters. We obtain a high-pass filter function by inverting the corresponding low-pass filter; for example, an ideal high-pass filter blocks all frequencies smaller than f_0 and leaves the others unchanged.

Bandpass filters are a combination of both low-pass and high-pass filters. They attenuate all frequencies smaller than frequency f_0 and higher than frequency f_1, whereas the frequencies between the two cutoffs remain in the resulting output image. We obtain the filter function of a bandpass filter by multiplying the filter functions of a low-pass and of a high-pass filter in the frequency domain, where the cutoff frequency of the low-pass filter is higher than that of the high-pass filter.

6.4 ALGORITHMIC ACCOUNT

The algorithmic descriptions of the various types of filters discussed earlier are given in the following subsections, divided according to the generic algorithms and not the specific filter type.

6.4.1 Spatial Filtering (Convolution Based)

In this type of filter design, a specific mask (or kernel) is needed. The nature of this mask will define the filtering type that will be applied to the image in the subsequent steps. Once this mask is initialized, the original image is first zero-padded or row-replicated to include additional rows and columns at the edges. This is done in accordance with the size of the mask

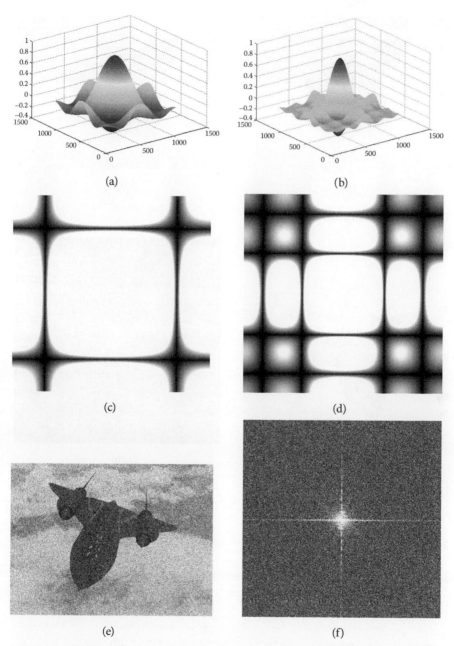

FIGURE 6.7 Filter masks and their effects. (a) Sharp cutoff circular mask; (b) Gaussian mask; (c) frequency domain representation of (a); (d) frequency domain representation of (b); (e) Gaussian-noise-corrupted image; (f) frequency domain representation of (e); (g) filtered image using (a); (h) filtered image using (b).

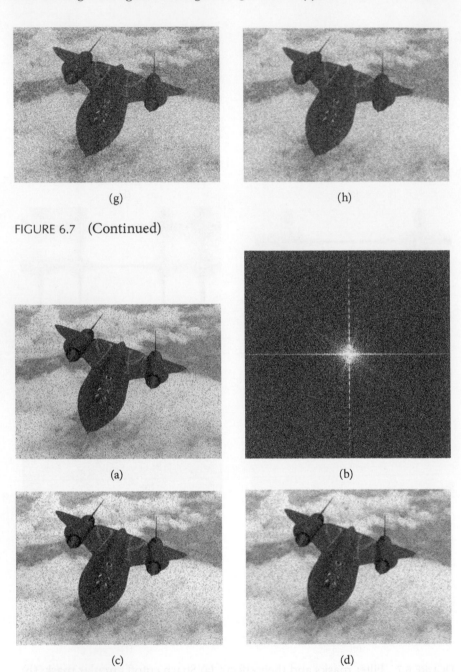

(g)   (h)

FIGURE 6.7   (Continued)

(a)   (b)

(c)   (d)

FIGURE 6.8   Filter masks and their effects on a salt-and-pepper-noise-corrupted image. (a) Corrupted image; (b) frequency domain representation of (a); (c) filtered image using sharp cutoff filter; (d) filtered image using a Gaussian filter.

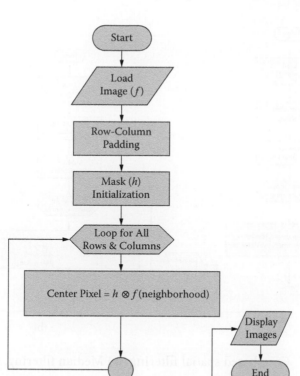

FIGURE 6.9 Kernel-based spatial filtering algorithm.

being used. If it is an odd order (say M × M), then there will be $\lfloor M/2 \rfloor$ pixels sticking out of the image in the form of a row as well as a column. Hence, having the additional rows and columns added first helps in avoiding numerical complications. Alternatively, the convolution process can be started with $\lfloor M/2 \rfloor$ pixels inside the image, leaving as many rows and columns on the boundary unprocessed. This is usually acceptable for a large enough image. After this preprocessing step, iterative convolution is performed according to Equation 6.2. The complete algorithm is shown in Figure 6.9.

6.4.2 Spatial Filtering (Case Based)

This category covers the general algorithmic idea that within a predefined masking area (not the mask!), the neighborhood of the central pixel is evaluated for certain cases. For instance, the median filter will replace the central pixel by the median of the sorted pixel list within this masking area. Conservative masking would look for two thresholds, minimum and

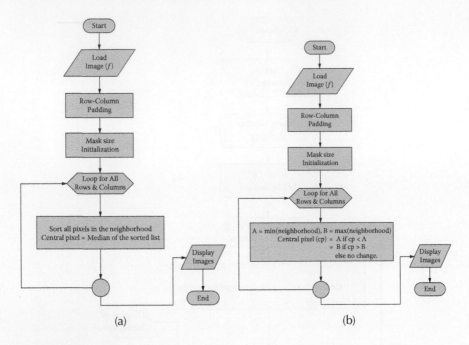

FIGURE 6.10 Case-based spatial filtering. (a) Median filtering, (b) conservative filtering.

maximum values of the eight neighbors of the central pixel, and decide on the replacement criteria. These two case-based filters can be thought of as the algorithms shown in Figure 6.10.

6.4.3 Frequency Filtering

Frequency filtering works very similarly to the spatial filtering shown in Figure 6.9, except for the image transformation steps in the beginning and end of the algorithm and replacement of convolution with multiplication. This is shown in Figure 6.11.

6.5 MATLAB® CODE

In this section, the MATLAB code used to generate the images in Figures 6.2, 6.3, 6.4, 6.6, 6.7, and 6.8 is presented. Except for Figures 6.7 and 6.8, all other listed figures are covered in the following code:

```
close all
clear all

h1 = ones(3)/9 ;
```

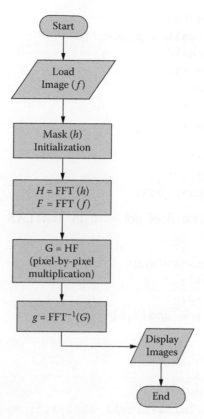

FIGURE 6.11 Frequency filtering algorithm.

```
h2 = fspecial('gaussian',5,5);
x=imread('sr71.jpg');
x = rgb2gray(x) ;
figure ; imshow(x)

% Gaussian Noise
xng = imnoise(x,'gaussian',0,0.1);
figure ; imshow(xng)
g1 = imfilter(xng,h1);
figure ; imshow(g1)
g2 = imfilter(xng,h2);
figure ; imshow(g2)
g3 = medfilt2(xng) ;
figure ; imshow(g3)
g4 = conserv(xng) ;
figure ; imshow(g4*0.003)
```

```
% Salt & Pepper Noise
xns = imnoise(x,'salt & pepper',0.1);
figure ; imshow(xns)
g5 = imfilter(xns,h1);
figure ; imshow(g5)
g6 = imfilter(xns,h2);
figure ; imshow(g6)
g7 = medfilt2(xns) ;
figure ; imshow(g7)
g8 = conserv(xng) ;
figure ; imshow(g8*0.003)
```

The function conserv does not exist in MATLAB and was coded as follows:

```
function y1 = conserv(xng)
[r,c] = size(xng) ;
x1 = zeros(r+2,c+2) ;
x1(2:r+1,2:c+1,:) = xng(:,:) ;
[r,c] = size(x1) ;
y1 = x1 ;

for i = 2 : r-1
    for j = 2 : c-1
            nh = [x1(i-1,j-1) x1(i-1,j) x1(i-1,j+1)
            x1(i,j-1) x1(i,j+1) x1(i+1,j-1) x1(i+1,j)
            x1(i+1,j+1)] ;
            cp = x1(i,j) ;
            mx = max(nh) ; mn = min(nh) ;
            if (cp > mx), cp = mx ;
            else if (cp < mn), cp = mn ;
                end
            end
            y1(i,j)= cp ;
    end
end
return
```

The following code is for Figures 6.7 and 6.8:

```
close all
clear all
```

```
h1 = ones(3)/9 ;
h2 = fspecial('gaussian',5,5);
H1 = fft2(h1, 1024,1024) ;
H2 = fft2(h2, 1024,1024) ;
figure ; mesh(fftshift(real(H1)))
figure ; mesh(fftshift(real(H2)))
figure ; imshow(fftshift(abs(H1))*20)
figure ; imshow(fftshift(abs(H2))*20)

x=imread('sr71.jpg');
x = rgb2gray(x) ;
[r,c] = size(x) ;

% Gaussian Noise
xng = imnoise(x,'gaussian',0,0.1);
figure ; imshow(xng)
XNG = fft2(xng,1024,1024) ;
figure ; imshow(fftshift(abs(XNG))*0.00001)
G1 = XNG .* H1 ;
g1 = ifft2(G1) ;
g1 = g1(1:r,2:c) ;
figure ; imshow(g1*0.004)
G2 = XNG .* H2 ;
g2 = ifft2(G2) ;
g2 = g2(1:r,2:c) ;
figure ; imshow(g2*0.004)

% Salt & Pepper Noise
xns = imnoise(x,'salt & pepper',0.1);
figure ; imshow(xns)
XNS = fft2(xns,1024,1024) ;
figure ; imshow(fftshift(abs(XNS))*0.00001)
G3 = XNS .* H1 ;
g3 = ifft2(G3) ;
g3 = g3(1:r,2:c) ;
figure ; imshow(g3*0.004)
G4 = XNS .* H2 ;
g4 = ifft2(G4) ;
g4 = g4(1:r,2:c) ;
figure ; imshow(g4*0.004)
```

6.6 SUMMARY

- Filtering is the operation of removing certain frequency components from an image.

- Image filters can work directly in the spatial domain as well as in the frequency domain.

- In the spatial domain, the filter can be thought of as a small M × N matrix, called *kernel* or *mask*, that is convolved with the source image.

- In such a convolution operation, a small neighborhood in the image (proportional to the kernel) is operated on, and a new pixel value is obtained as the corresponding center pixel of the neighborhood.

- For the case of boundaries, additional rows and columns are added that are either replicas of the outer rows and columns or just zeros (zero padding).

- The spatial filters can also be based on checking certain constraints within the neighborhood, so that outlier values can be smoothed out. Conservative and median filtering are two such filters.

- The images can also be processed in the frequency domain. Here, a specific mask is needed to remove or allow certain areas of the frequency spectrum.

- Noise is usually the higher frequency aspect of the image and can be removed by the low-pass filter mask, which is usually some type of circular and zero-centered mask.

- However, removing the noise frequencies also removes with them some useful high-frequency components such as edges in the image, thus producing a blurred image.

- Careful selection of the mask and its application can achieve a compromise solution in which noise removal and edge conservation are optimized.

6.7 EXERCISES

1. Apply a high-pass filter to the image in Figure 6.7(e). Reduce the power of the Gaussian noise by a factor of 4, and repeat the high-pass filtering for this less noisy image. Comment on the edge enhancement in the image in both cases. [Hint: Check the help

documentation of the `fspecial()` function in MATLAB for the design of high-pass filters]

2. Apply a conservative filter to the image in Figure 6.6 by reducing the salt-and-pepper noise by a factor of 4. Then compare with the results in Figure 6.6.

3. Take a heavily noisy image, for example, Figure 6.3(a). Apply the Gaussian spatial filter of size 7 × 7. Then apply the filter obtained in Q #1 to the filtered image. Is this iterative procedure better than simple low-pass filtering? Can the iterative procedure be continued further? If so, for typically how many more times before losing significant information?

4. Referring to the procedure for Figure 6.8, perform the following experiment:

   a. Once the 1024 × 1024 frequency domain representation of the noisy image is obtained, use the mask shown in Figure EX6.1(a) as is to remove certain portions of the frequency domain image.

   b. Inverse-transform the resulting image to obtain the filtered spatial domain representation of the filtered image.

5. Repeat Q #4 for the mask shown in Figure EX6.1(b).

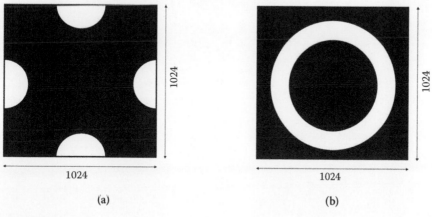

(a)   (b)

FIGURE EX6.1

FIGURE E6.X

CHAPTER 7

Image Restoration and Blind Deconvolution

7.1 INTRODUCTION

IMAGE ENHANCEMENT, IN GENERAL, covers all aspects of improving image quality for visual purposes. This includes

1. Noise removal

2. Contrast stretching

3. Histogram equalization

4. Filtering

5. Restoration through deconvolution

All convolution-related topics have been covered in earlier chapters except image restoration and the underlying mathematics of deconvolution; these will be discussed in this chapter.

Images are produced in order to record or display useful information. No image information is perfect because of inherent physical and technological limitations. With the ever-increasing demand for improvement in the quality of digital images, there is a need to correct errors caused by the imaging system. These errors have many causes, including the hardware; insertion of additive or multiplicative noise by both human and other sources; and blurring of the image by linear motion of the imaging equipment or misalignment, causing defocusing.

One of the most intriguing questions in image processing is the problem of recovering the desired or perfect image from a degraded version. The amount of lost information that can be recovered from the blurred version of an image depends on the extent of degradation. This sets a bound on the performance of the image restoration algorithm.

In order to solve the image restoration problem, mathematical models are required for real-world processes involved in image generation, formation, and recording. It may be argued that the ultimate objective of a restoration scheme is the inverse manipulation of the degraded model into the original nondegraded model. If this task is performed based on certain statistical knowledge, known beforehand, about the degradation process, the medium, and the noise, the procedure is called *a priori restoration*. On the other hand, this information may not be available with a high degree of fidelity. In that case, *blind restoration* or blind deconvolution methodologies are adopted.

7.2 IMAGE REPRESENTATION

In order to understand the functioning of certain restoration techniques, the morphology of the target image and the degradation in it should be understood first. The general notation for digital images is to consider them as matrices represented in x and y coordinates. Thus, an image of interest $f(m,n)$ is a matrix of $m \times n$ order that is the result of 2D sampling of the real image. However, the image model $f(m,n)$ is not the observed image because of the distortion present in the imaging system. A general imaging system block diagram is shown in Figure 7.1.

Hence, the observed image $g(m,n)$ is given by

$$g(m,n) = f(m,n) \otimes h(m,n) + \eta(m,n), \qquad (7.1)$$

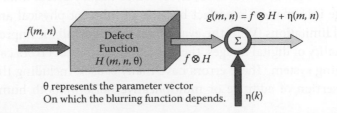

FIGURE 7.1 Imaging system and its block diagram.

where $h(m,n)$ represents the degradation model, and $\otimes$ represents the 2D convolution operation. The additive term $\eta(m,n)$ represents the noise added to the degraded image, further increasing the distortion of image. The blurring function in the main block is a function of the pixel coordinates m and n, and a parameter list θ. This list depends on the type of representation used for the blurring function. For instance, for a Gaussian function, the parameter list only comprises two parameters, mean and variance, and so on. In some applications, noise can also be considered multiplicative, in which case the restoration problem is more difficult than in the additive noise case. The function $h(m,n)$ is also called the *point spread function* (PSF) in the image processing literature, and depends on the type of distortion or degradation.

The degradation $h(m,n)$ can be convolved with the clean image because of several factors. Among them, the most prominent are hardware misalignment, malfunction, human errors, and electronic noise. In nearly any practical situation, a device of finite spatial extent records an image. As a result, not all the data that is needed to model the real object is available.

Convolution in the Fourier domain is identical in result to that in the spatial domain, in which a kernel of values (a blur function) is applied to every pixel in the image. In Fourier space, this simply involves multiplying the transform of the blur function with the transform of the image, pixel by pixel. In many cases, it will be easier to construct the kernel as an image rather than as a set of numerical values. This is particularly true when the kernel is not symmetrical. Sometimes, the blur function can be measured directly by acquiring an image of a point (e.g., a star in astronomical images). Some examples of such convolutions are shown applied to the car image in Figure 7.2.

This also implies that it is possible to remove blur from a blurred image by dividing by the transform of the blur, and this process is called *deconvolution*. It deconvolves the main image using the content of the image previously designated. Deconvolution is typically used to remove blur from an image due to out-of-focus optics or motion while capturing the image with a camera. There are other applications in which deconvolution can be employed, to remove specific blurring function. In general, when deconvolving such blurs, one image is the blurred image and a second image is used that contains information related to the PSF of the blur only. By using this second image in the same format as that of the blurred image, it facilitates the algorithmic procedures in terms of speed and memory requirements.

Figure 7.2 shows what it means to perform deconvolution. The car2.jpg image has first been convolved with two different types of blur or PSFs and

(a)

(b)

(c)

(d)

FIGURE 7.2 Example of deconvolution for two commonly occurring blurring problems. (a) Motion-blurred image, (b) deconvolved image, (c) out-of-focus blurred image, (d) deconvolved image.

then recovered correspondingly through the inverse process of deconvolution. The first example is a symmetrical blur as might be produced by out-of-focus optics, and the second is motion blur, which corresponds to a straight line (whose direction and distance must be known).

When the PSF can be measured from the image itself, for example, as the blurred image of a single star, a single fluorescent bead, or a single dust particle, it can be conveniently used for deconvolution. In other cases, it may be possible to estimate it from independent knowledge.

Frequency filters are also commonly used in image reconstruction. Here, the aim is to remove the effects of a nonideal imaging system by multiplying the image in the Fourier space with an appropriate function. The easiest method, called *inverse filtering*, is to divide the image in the Fourier space by the optical transfer function (OTF). We illustrate this deconvolution, using Figures 7.3 and 7.4.

7.3 DECONVOLUTION

Deconvolution is the reverse (at least theoretically) of convolution. When the degradation due to blur is to be removed, the process is not as simple as filtering out the noise. Because the blurring of the image is essentially a convolution of the clean image with the PSF of the blur, the principal procedure in any restoration scheme is finding out an inverse procedure to cancel out the effect of the blur PSF. This is also called deconvolution, or inverse filtering. Given the image model of Equation 7.1 without noise, then

$$g = h \otimes f \quad \overset{\Im}{\Leftrightarrow} \quad G = HF, \tag{7.2}$$

where G, H, and F are the Fourier transforms for g, h, and f, respectively. Therefore, by inverse filtering or a deconvolution operation, the original image can be retrieved as follows:

$$F = D^{-1}G \quad \overset{\Im}{\Leftrightarrow} \quad f = (\Im^{-1}h^{-1}) \otimes g, \tag{7.3}$$

where $\Im^{-1}$ represents the inverse Fourier transform. This seemingly simple problem is not so simple in reality. In many practical cases, it is useless or even impossible to apply it as given by Equations 7.2 and 7.3. This is because the PSF is usually unknown and often zero over wide ranges. Hence, h^{-1} will be infinite in this case.

If additive noise is considered for the inverse filtering problem, then the end result will be similar to Equation 7.3 but with an additional term, $h^{-1}\eta$. This means that even if H is nonzero, the noise will be amplified by a constant factor of h^{-1}. In effect, the signal-to-noise ratio in this case is not improved but stays the same because noise and the degraded system are amplified by the same factor.

There are three situations with respect to the blurring function and its know-how:

White box problem: When *h* is precisely known

Gray box problem: When partial information is available about *h*

Black box problem: When nothing is known about the blurring function

When the blurring function $h(m,n)$ is not accurately known, it must be estimated prior to inverse filtering. Because the attempt is to deconvolve the degraded image $g(m,n)$ without any prior knowledge of the cause that is producing the degradation, such a procedure is also called *blind deconvolution*, as it is blind to the source of the degradation.

In MATLAB®, three main types of deconvolution algorithms have been implemented. This is by no means an exhaustive list, and newer methods keep popping up in the literature from time to time. The chief methods are

1. Lucy–Richardson Method

2. Wiener method

3. Blind method

The algorithms related to these methods are presented in Section 7.4. Figure 7.3 shows the results of using the Lucy–Richardson, Wiener, and blind deconvolution methods on the car2.jpg image. Specifically, for blind deconvolution, two initial blurring functions were tested. The first is a random matrix of size 5 × 5, and the second is the same blurring function that caused the blurring in the first place.

The same example is repeated in Figure 7.4 for the motion blur case, with motion affecting a neighborhood of 15 pixels and at 45°.

In general, these algorithms either use a known (or somewhat known) blurring function or try to estimate one iteratively. Usually, they rely on second-order statics or H_2 norms for their convergence. Optimal H_2 filters (or Kalman filters) are suitable solutions to the filtering and estimation problem when the power spectral density of the noise is precisely known. In general least-squares estimation problems, one minimizes the integral of the power spectral density estimation error. This minimization of the average error power or error variance might result in a relatively large error power in some frequency range. In many practical situations, however,

FIGURE 7.3 Results of deconvolution with out-of-focus blur. (a) original image, (b) blurred image, (c) deconvolved image with Lucy–Richardson method, (d) deconvolved image with Wiener method, (e) deconvolved image with blind method using random initial blurring function, (f) deconvolved image with blind method using the same initial blurring function.

there is significant uncertainty in the power spectral density of the noise. An appropriate criterion, as argued by Zames, is the H_∞ norm of the estimation error spectrum. This H_∞ problem based on the minimization of the magnitude of the estimation error spectrum rather than the average power leads to better filters compared to other solutions. The H_∞ optimization scheme leads to improved performance of the deconvolution process over existing least-squares or H_2-based solutions, thus leading to enhanced defect impulse response and thereby improving the defect identification.

Given a linear time-invariant (LTI) system with input $f(t)$, output $g(t)$, and a transfer function $H(s)$, the H_∞ norm is expressed in time domain as

$$\| H \|_\infty^2 = \max_{\|f\|_2 \leq 1} \left\{ \frac{\| g \|_2^2}{\| f \|_2^2} \right\} = \max_{\|f\|_2 \leq 1} \left\{ \frac{\sum g^2(k)}{\sum f^2(k)} \right\}. \tag{7.4}$$

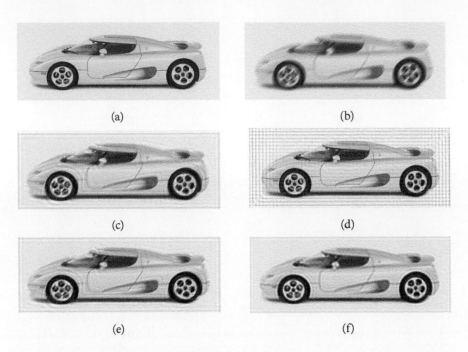

(a) (b)

(c) (d)

(e) (f)

FIGURE 7.4 Results of deconvolution with motion blur. (a) Original image, (b) blurred image, (c) deconvolved image with Lucy–Richardson method, (d) deconvolved image with Wiener method, (e) deconvolved image with blind method using random initial blurring function, (f) deconvolved image with blind method using the same initial blurring function.

Therefore, the H_∞ norm is interpreted as the maximum energy gain of the system for all finite energy signals. In the frequency domain, the H_∞ norm is expressed as

$$\| H \|_\infty = \max_\omega | H(j\omega)|, \qquad (7.5)$$

that is, it is the largest peak of the frequency response magnitude.

Once the input–output model is formulated in the form of a transfer function, it can be related to an autoregressive moving average (ARMA) process as follows:

$$D(z^{-1})g(k) = N(z^{-1})f(k) + n(k), \qquad (7.6)$$

where $g(k)$ is the output of the process, $f(k)$ is the input to the process, $n(k)$ is the driving noise, and

$$D(z^{-1}) = 1 + a_1 z^{-1} + a_2 z^{-2} \cdots + a_{na}z^{-na}, \qquad (7.7)$$

$$N(z^{-1}) = b_0 + b_1 z^{-1} + \cdots + b_{nb} z^{-nb}, \tag{7.8}$$

where n_a and n_b are the degrees of the polynomials $D(z^{-1})$ and $N(z^{-1})$, which are the denominator and numerator of the ARMA model, respectively. The model (Equations 7.6–7.8) is now put in a more suitable form for application of an identification method.

$$g_k = \Psi_k^T \theta + n_k, \tag{7.9}$$

where g_k denotes the samples of the output,

$$\Psi_k^T = [-g(k{-}1) \ {-}g(k{-}2) \ \cdots \ {-}g(k{-}na) \ f(k) \ f(k{-}1) \ \cdots \ f(k{-}nb)] \tag{7.10}$$

and

$$\theta^T = [\, a_1 \ a_2 \ \cdots \ a_{na} \ b_0 \ b_1 \ \cdots \ b_{nb}]. \tag{7.11}$$

A linear parameter estimator E must be found to operate upon such that the unbiased estimate at $i = k$ is denoted by

$$\hat{\theta}_k = E(g_i, f_i, i \le k). \tag{7.12}$$

The estimator E should be chosen for the worst possible n_k (representing noise and modeling errors) that produces a best estimate of the original image in the following sense. Denote

$$e_\gamma = \gamma^{-1}(\theta - \hat{\theta}). \tag{7.13}$$

Hence, the following differential-game problem can be minimized for a value of γ that is as close as possible to the minimum possible value γ_o:

$$\sup_{n \in L_2[0 \cdots N]} \frac{\|\theta - \hat{\theta}\|_2^2}{\|n_k\|_2^2 + \|\theta\|_{R_o^{-1}}^2} < \gamma^2. \tag{7.14}$$

Figure 7.8 describes an algorithm for implementing this methodology.

7.4 ALGORITHMIC ACCOUNT

There are several methods available in the literature to perform the deconvolution operations on certain classes of distorted images. However, this section deals with only those techniques that are part of MATLAB's toolbox.

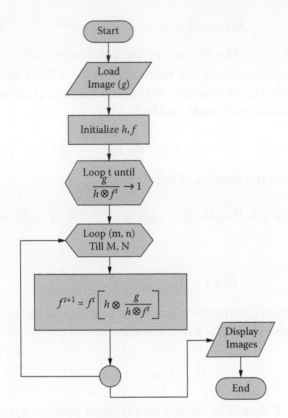

FIGURE 7.5 The Lucy–Richardson algorithm.

7.4.1 Lucy–Richardson Method

This method is a classical technique developed in the early to mid 1970s and was used extensively for restoring astronomical images. In this statistical technique, the same model is used as presented in Equation 7.1. However, additive noise from the model is ignored due to the preprocessing step containing low-pass filtering. The algorithm utilizes a known blurring function and iterates with it until it converges. The algorithm is shown in Figure 7.5.

7.4.2 Wiener Method

This is a classical regularization algorithm that introduces additional information about the problem in order to obtain a well-behaved inverse. Basically, the procedure tries to make the convolution model Equation 7.1 a well-posed system, which can be described as one that is *uniquely solvable* and is such that the *solution depends in a continuous way on the data*. Also, if

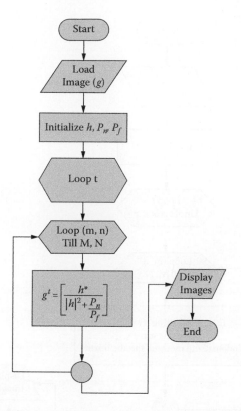

FIGURE 7.6   Algorithm for Wiener method of deconvolution.

the solution depends in a discontinuous way on the data, then small errors, whether rounding-off errors, measurement errors, or perturbations caused by noise, can create large deviations. The algorithm incorporates prior knowledge (or assumptions) about the statistical properties of the image, such as the following: intensity is always positive; image is the expected form of the signal; pixels are generated by point sources (also called CLEAN algorithm), signal smoothness, and most importantly, expected signal and noise power spectra. The algorithm is shown in Figure 7.6.

7.4.3 Blind Deconvolution

The foregoing two methods usually rely on the known a priori information related to the blurring function h. This can be a serious handicap in many industrial as well as scientific applications. Hence, for a black-box problem, the inverse approach is called blind deconvolution. Usually, a Kalman-type state estimator is used in the design. The idea is to initialize

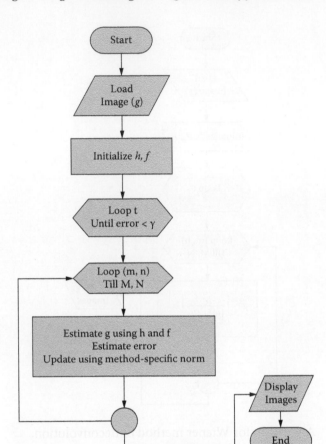

FIGURE 7.7 General blind deconvolution algorithm.

an iterative algorithm with a preselected structure (state-space or transfer function) with random or flat initialization. Using this assumed structure, a new image estimate is then obtained. A tuning parameter is then adjusted based on the error between this estimate and the actual image. Because nothing is known about the true image as well, some statistical assumptions are needed to decide on the correction. Usually, some form of covariance matrix is used to adjust the underlying model, and then the iterative procedure is continued until a suitable convergence is reached. The general algorithm is shown in Figure 7.7.

As discussed toward the end of Section 7.3, the H_2 norm-based solution has certain statistical limitations, and H_∞ can be a better approach.

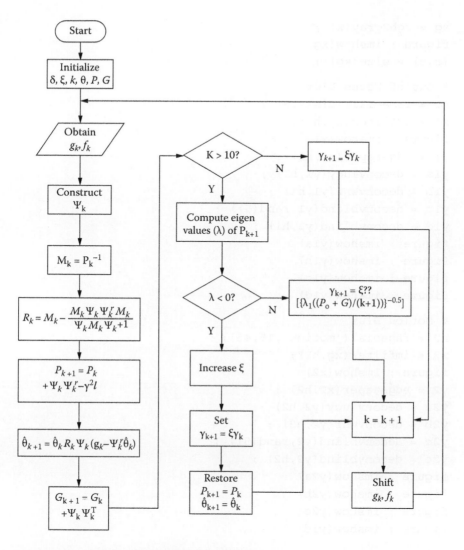

FIGURE 7.8 H_∞ deconvolution algorithm.

Figure 7.8 depicts one such algorithm, which was originally designed for 1D signals but later on was modified extensively for 2D applications.

7.5 MATLAB CODE

In this section, the MATLAB code used to generate the images in Figures 7.2, 7.3, and 7.4 is presented:

```
close all ; clear all
x=imread('car2.jpg');
```

```
xg = rgb2gray(x) ;
figure ; imshow(xg)
[r,c] = size(xg) ;

% Out of Focus Blur
h1 = fspecial('disk',5) ;
x1 = imfilter(xg,h1);
figure ; imshow(x1)
y1 = edgetaper(x1,h1) ;
y1a = deconvlucy(y1,h1) ;
y1b = deconvwnr(y1,h1) ;
y1c = deconvblind(y1,rand(5)) ;
y1d = deconvblind(y1,h1) ;
figure ; imshow(y1a)
figure ; imshow(y1b)
figure ; imshow(y1c)
figure ; imshow(y1d)

% Motion Blur
h2 = fspecial('motion',15,45);
x2 = imfilter(xg,h2);
figure ; imshow(x2)
y2 = edgetaper(x2,h2) ;
y2a = deconvlucy(y2,h2) ;
y2b = deconvwnr(y2,h2) ;
y2c = deconvblind(y2,rand(5)) ;
y2c = deconvblind(y2,h2) ;
figure ; imshow(y2a)
figure ; imshow(y2b)
figure ; imshow(y2c)
figure ; imshow(y1d)
```

7.6 SUMMARY

- Image restoration refers to image improvement against deformations such as blur.

- The blurring process can be modeled as a mathematical convolution operation between the true image and the blurring function.

- If the blurring function is known, the problem is called a *white box problem*.

- If the blurring function is partially known, the problem is called a *gray box problem.*

- If the blurring function is completely unknown, the problem is called a *black box problem.*

- The usual methods of deconvolution rely on the availability of the information related to the blurring function.

- These methods usually rely on the application of the inverse of the Fourier transform of the blurring function to the Fourier transform of the blurred image.

- Blind deconvolution is performed when there is a black box problem in hand.

- Here, usually an initial estimate of the true image as well as the blurring function are picked up randomly and iteratively evolved to the blurred image, which then gives a better estimate of the blurring function.

- After this, the usual frequency domain deconvolution can be applied to get the estimate of the true image.

- Deconvolution is not a very simple process, and usually the visual quality cannot match the true image. However, a lot of improvement can be obtained.

7.7 EXERCISES

1. Regenerate the image in Figure 7.3(b), which is blurred owing to out-of-focus blur. Then use the motion-blurring function as the initial blurring function with the following, and compare the results with Figures 7.3(c), (e), and (f):

    a. Lucy–Richardson method.

    b. Blind deconvolution

2. Repeat Q #1 for Figure 7.4(b), using the out-of-focus blurring function.

3. As the Wiener function is optimal with respect to Gaussian noise, can the same be said for the deconvolution process? Explain your answer by comparing the results of deconvolution using the Wiener

method on three images with out-of-focus, Gaussian, and motion blurs, respectively.

4. An interesting method of removing out-of-focus blur could be repeated application of a low-pass filter and then a high-pass filter. Demonstrate the feasibility or otherwise of this technique. Also comment on a useful count for the number of iterations.

5. Another variant of the procedure in Q #4 is to first iteratively apply high-pass filtering, and then use histogram equalization for intensity distribution. Compare this procedure with the one in Q #4.

CHAPTER 8

Image Compression

8.1 INTRODUCTION

EACH PIXEL IN AN image is represented by a certain number of bits (as supported by the system). It is interesting to calculate how much memory is needed to store such an image. For example, consider the image sbj. bmp, which is an RGB image of dimensions 340 × 470 and requires 468K bytes for its representation and storage in this form. This number is obtained through the calculation (340 × 470 × 3)/1024. Since most Windows systems consider the byte to be 8 bits long, each RGB triad requires 3 bytes for its representation. As such, the storage requirements for RGB images are three times those for grayscale images. However, the foregoing image requires only 12K bytes when it is stored as a JPEG image! This is the domain of image compression, in which the intent is to design algorithmic techniques that can reduce the storage requirements of the image and, at the same time, retain the image information content. There are several aspects of image properties that can be exploited in the process of compression. Distinctively, the interpixel information variation is only significant at edges of any type, whereas most of the image information content remains a slowly changing variable. Similarly, the fact that our eyes are less sensitive to color changes and are much more sensitive to intensity variations indicates that color space can be compromised more than grayscale intensity space. Representation of color space differently can also help in reducing the image size for storage. It should be noted that compression is realized only when the image is stored on the disk. Once it is loaded in memory, the image occupies all the required RAM locations according to the size and color requirements for the original BMP-type structure.

Compressing an image is significantly different from compressing raw binary data. Of course, general-purpose compression programs can be used to compress images, but the result is less than optimal. This is because images have certain statistical properties that can be exploited by encoders specifically designed for them. Also, some of the finer details in the image can be sacrificed for the sake of saving a little more bandwidth or storage space. This also means that lossy compression techniques can be used in this area. Lossless compression deals with compressing data in such a way that an exact replica of the original data is obtained upon decompression. This is the case when binary data such as executables, documents, etc., are compressed. They need to be exactly reproduced when decompressed. On the other hand, images need not be reproduced exactly. An approximation of the original image is enough for most purposes, as long as the error between the original and the compressed image is tolerable.

In this chapter, we will take a close look at compressing grayscale images. The algorithms explained can be easily extended to color images, either by processing each of the color planes separately, or by transforming the image from RGB representation to other convenient representations such as YUV or YC_bC_r in which processing is much easier.

8.2 IMAGE COMPRESSION–DECOMPRESSION STEPS

The usual steps involved in compressing an image are

1. **Specification:** This implies specifying the rate (bits available) and distortion (tolerable error) parameters for the target image.

2. **Classification:** This implies dividing the image data into various classes, based on their importance. Usually, some type of compression transform is utilized in this step to associate the important features with the most important class of information to be kept in the process of compression.

3. **Bit allocation:** This implies dividing the available bit budget among these classes such that the distortion is a minimum.

4. **Quantization:** This refers to quantizing each class separately using the bit allocation information derived in step 3.

5. **Encoding:** This corresponds to encoding each class separately using an entropy coder and write to the file.

Most of the image compression techniques work on these steps. However, there are exceptions, as well those that are designed for specific applications. One example is the fractal image compression technique, where possible self-similarity within the image is identified and used to reduce the amount of data required to reproduce the image. Traditionally, these methods have been time consuming, but some latest methods promise to speed up the process.

Decompressing or reconstructing the image from the compressed data is usually a faster process than compression. The steps involved are

1. **Decoding:** Read in the quantized data from the file, using an entropy decoder (reverse of step 5).

2. **Dequantizing:** This refers to normalizing the quantized values (reverse of steps 4 and 3). This also includes any padding or addition of missing values due to the quantization process.

3. **Rebuilding:** This involves the inverse transform or inverse classification of the normalized data into image pixels, essentially rebuilding the image (reverse of step 2).

8.2.1 Error Metrics

The fundamental question of how compression techniques can be compared is quantitatively answered by using two error metrics: the mean square error (MSE) and the peak signal-to-noise ratio (PSNR). The MSE is the cumulative squared error between the compressed and the original image, whereas the PSNR is a measure of the peak error. The mathematical formulae for the two are

$$MSE = \frac{1}{M \times N} \sum_{m=1}^{M} \sum_{n=1}^{N} [f(m,n) - g(m,n)]^2, \tag{8.1}$$

$$PSNR = 20 \times \log_{10}\left(\frac{255}{\sqrt{MSE}}\right), \tag{8.2}$$

where $f(m,n)$ is the original image, $g(m,n)$ is the approximated version (which is actually the decompressed image), and M,N are the dimensions of the images. A lower value for MSE means less error, and as is seen from the inverse relation between the MSE and PSNR, this translates to a high value of PSNR. Logically, a higher value of PSNR is good because it means that the ratio of signal to noise is higher. Here, the signal is the original image, and the noise is the error in reconstruction.

8.3 CLASSIFYING IMAGE DATA

In this step, usually the image is represented as a two-dimensional array of coefficients, where each coefficient represents the brightness level at that point. From a high-level perspective, one cannot differentiate between coefficients as more important ones and lesser important ones. But it is intuitively possible to do so. Most natural images have smooth color variations, with the fine details being represented as sharp edges between the smooth variations. Technically, the smooth variations in color can be termed low-frequency variations and the sharp variations, high-frequency variations.

The low-frequency components (smooth variations) constitute the base of an image, and the high-frequency components (the edges that give the detail) add upon them to refine the image, thereby giving a detailed image. Hence, the smooth variations require more importance than the details.

Separating the smooth variations and details of the image can be done in many ways. Two well-known image transforms used for this purpose are the discrete cosine transform (DCT) and the discrete wavelet transform (DWT). The DCT is discussed in the next section, and the DWT was discussed in Chapter 5.

8.3.1 Discrete Cosine Transform

Another sinusoidal transform (i.e., transform with sinusoidal base functions) related to the DFT is the DCT. For an N × N image, the DCT is given by

$$F(p,q) = \frac{1}{4}\psi(p)\xi(q)\sum_{m=0}^{N-1}\sum_{n=0}^{N-1}f(m,n)\cos\left(\frac{(2m+1)p\pi}{2N}\right)\cos\left(\frac{(2n+1)q\pi}{2N}\right)$$

(8.3)

where $f(m,n)$ is the original image in the spatial domain with indices m and n, $F(p, q)$ is the transformed image with indices p and q, and the transformation coefficient multipliers are given by Ψ and ξ, which are defined as

$$\psi(p) = \begin{cases} \dfrac{1}{\sqrt{2}} & for \quad p=0 \\ 1 & otherwise \end{cases}$$

(8.4)

$$\xi(q) = \begin{cases} \dfrac{1}{\sqrt{2}} & for \quad q = 0 \\ \\ 1 & otherwise \end{cases}.$$

(8.5)

The main advantages of the DCT are that it yields a real-valued output image and that it is a fast transform. After performing a DCT, it is possible to throw away the coefficients that encode those frequency components that the human eye is not very sensitive to. Thus, the amount of data can be reduced, without seriously affecting the way an image looks to the human eye. The inverse transform for Equation 8.3 is given by

$$f(m,n) = \frac{1}{4} \sum_{p=0}^{N-1} \sum_{q=0}^{N-1} \psi(p)\xi(q)F(p,q)\cos\left(\frac{(2m+1)p\pi}{2N}\right)\cos\left(\frac{(2n+1)q\pi}{2N}\right).$$

(8.6)

The definition of the coefficient multipliers remains the same as in Equations 8.4 and 8.5. Figure 8.1 shows the application of DCT and inverse DCT to the cameraman image.

8.4 BIT ALLOCATION

The first step in compressing an image is to segregate the image data into different classes. Depending on the importance of the data it contains, each class is allocated a portion of the total bit budget such that the compressed image has the minimum possible distortion. This procedure is called *bit allocation*.

The rate-distortion theory is often used for solving the problem of allocating bits to a set of classes, or for bit-rate control in general. The theory aims at reducing the distortion for a given target bit rate by optimally allocating bits to the various classes of data. One approach to solving the problem of optimal bit allocation is through using the rate-distortion theory explained as follows:

1. Initially, all classes are allocated a predefined maximum number of bits.

2. For each class, one bit is reduced from its quota of allocated bits, and the distortion due to the reduction of that one bit is calculated.

(a)   (b)

(c)

FIGURE 8.1 Application of the discrete cosine transform (DCT) and inverse DCT. (a) Original image, (b) DCT of the whole image, (c) recovered image using inverse DCT.

3. Of all the classes, the class with minimum distortion for a reduction of one bit is noted, and one bit is reduced from its quota of bits.

4. The total distortion for all classes D is calculated.

5. The total rate for all the classes is calculated as $R = P(i) \times B(i)$, where P is the probability and B is the bit allocation for each class.

6. Compare the target rate and distortion specifications with the values obtained in steps 4 and 5. If not optimal, go to step 2.

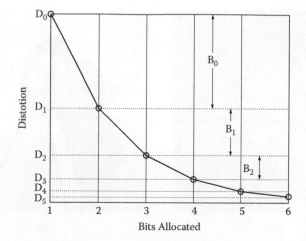

FIGURE 8.2 "Benefit" of a bit is the decrease in distortion due to receiving that bit.

In this approach, we keep reducing one bit at a time until we achieve optimality either in distortion or target rate, or both. An alternate approach is to initially start with zero bits allocated for all classes, and then to find the class that is most "benefited" by getting an additional bit. The benefit for a class is defined as the decrease in distortion for that class. Typically, it is defined by a decreasing parabolic curve, as shown in Figure 8.2.

As shown earlier, the benefit of a bit is a decreasing function of the number of bits allocated previously to the same class. Both approaches mentioned earlier can be used for the bit allocation problem.

Consider the same image as in Figure 8.1(a); this time we do not transform the image as a whole. Rather, the image is divided up into smaller blocks of a predefined size, and the DCT is calculated for each block independently. Each DCT block is then reverse-transformed to obtain the original image. This is shown in Figure 8.3.

8.5 QUANTIZATION

Quantization refers to the process of approximating the continuous set of values in the image data with a finite (preferably a much smaller) set of values. The input to a quantizer is the original data, and the output is always one among a finite number of levels. The quantizer is a function whose output values are discrete, and usually finite. Obviously, this is a process

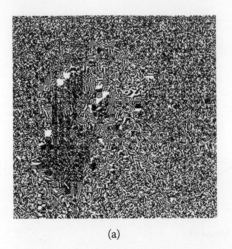

(a)   (b)

FIGURE 8.3   The DCT applied to blocks of image. (a) Application of DCT on isolated blocks, (b) recovered image.

of approximation, and a good quantizer is one that represents the original signal with minimum loss or distortion.

There are two types of quantization, scalar quantization and vector quantization. In scalar quantization, each input symbol is treated separately in producing the output, whereas in vector quantization, the input symbols are clubbed together in groups called *vectors*, and processed to give the output. This clubbing of data and treating them as a single unit increases the optimality of the vector quantizer, but at the cost of increased computational complexity.

Repeating the procedure depicted in Figure 8.3 with a bit of quantization can reveal more interesting facts about the compression involved. Actually, so far the DCT has produced another image of the same size! Note from Figure 8.1(b) as well as from segments in Figure 8.3(a) that the DCT has most of its significant coefficients clustered toward the top-left corner of the matrix. For example, Figure 8.4 shows some of these matrix blocks from Figure 8.3(a).

Assume that 4 out of 8 rows and columns in each block are now made zero. This would be a severe compression and may be avoided by using less depth or by using numbers closer to the values in these rows and columns rather than zeros. The resulting counterparts of the matrices in Figure 8.4 are shown in Figure 8.5. The interesting feature that can be noticed is that because half of the block is made up of all zeros, they become redundant. This implies that we need not store or transmit these

	1	2	3	4	5	6	7	8
1	1254.875	-1.9711	0.6581	0.4709	2.375	1.0406	2.5687	-0.4977
2	3.8203	-0.7165	0.528	0.7221	-1.9537	0.2029	-2.2978	4.0017
3	2.0439	0.5134	3.365	-1.4406	-2.3705	-1.5856	-0.4848	-0.6546
4	-4.9979	0.0819	-3.1677	-0.4627	0.911	-0.909	0.5341	-0.2885
5	3.375	-0.4348	0.3218	0.6203	0.875	0.2899	-2.1628	0.5289
6	-2.079	0.9224	1.1297	-1.4267	-2.559	0.0485	0.3766	1.4228
7	1.9947	-0.728	1.0152	-1.3862	-2.13	-5.4382	-2.115	-0.7537
8	-2.8577	0.9839	-1.2675	-2.0439	-1.2717	-0.4118	-3.8042	3.1307

(a)

	65	66	67	68	69	70	71	72
65	841	302.1462	104.7651	30.5575	14	1.2765	2.0653	-3.71
66	456.0342	-108.9023	-130.2682	-72.406	-26.4795	-12.297	-1.2748	-2.4157
67	66.1055	-185.9688	4.5585	70.4124	50.5773	25.3067	9.7522	3.9413
68	-9.0434	1.4466	82.8067	14.0929	-40.3098	-36.1525	-22.282	-5.1615
69	-53.5	36.4496	0.9816	-48.9929	-18.5	25.8295	27.1944	17.2489
70	-18.8813	43.8048	-19.3503	-6.3284	22.7722	15.804	-7.3555	-15.91
71	-4.3809	6.0429	-24.2478	16.6677	9.0866	-17.6497	-12.0585	-7.9933
72	8.8122	2.4084	-10.0236	16.1536	-4.5833	-12.5494	17.4176	20.0054
73	127.625	14.7686	3.4461	5.8781	4.875	3.6376	2.7668	0.8894

(b)

	129	130	131	132	133	134	135	136
129	873.375	-697.752	105.665	94.4273	-58.125	55.9427	-9.8078	-69.514
130	-9.4687	12.7594	13.1746	-6.7241	-12.3344	9.65	-1.1011	-10.2776
131	-8.2965	-1.3019	-3.4205	3.2375	3.1591	8.2861	4.3865	7.6099
132	-0.8682	-3.3157	-5.7457	4.4177	5.0785	2.4324	4.3479	5.5348
133	-2.125	3.1865	1.2225	-2.0555	-2.125	-0.7872	-0.6417	-1.3925
134	3.5626	2.3313	1.4306	-0.8428	-2.8238	-3.0746	-3.3014	-1.778
135	-0.7577	2.7431	0.3865	0.1223	2.0739	-3.4376	-1.8295	3.0859
136	-2.3367	-2.0527	-0.6625	-0.6876	1.2878	2.9097	1.2413	-0.1025

(c)

	249	250	251	252	253	254	255	256
249	917.375	-20.4205	-35.3913	0.9938	-20.875	22.5457	-10.6414	-14.7104
250	-17.4023	41.9748	13.6701	4.1349	8.2	-3.1763	-20.444	-16.5316
251	52.8833	22.1493	-26.8932	-1.3844	-4.0842	16.7127	12.1564	10.6299
252	52.9053	-19.3133	-22.4387	26.6849	9.1569	9.4899	-3.4021	13.551
253	9.625	-3.1002	31.3595	24.3583	-1.125	16.7293	7.4406	14.5476
254	-14.8952	-3.4955	13.5986	3.7178	3.9666	19.7072	1.8681	10.6729
255	-13.1105	-11.8632	-18.3436	-10.7882	17.2511	15.6234	-5.8568	-12.5538
256	-4.9628	14.6963	7.898	-18.8098	2.6907	18.0795	17.4838	-23.8669

(d)

FIGURE 8.4 8 × 8 matrix blocks from four different places in the image of Figure 8.3(a). White cells show the highest value in the block, and the top row and leftmost column show the column and row numbers, respectively.

	1	2	3	4	5	6	7	8
1	1254.875	-1.9711	0.6581	0.4709	2.375	1.0406	2.5687	-0.4977
2	3.8203	-0.7165	0.528	0.7221	-1.9537	0.2029	-2.2978	4.0017
3	2.0439	0.5134	3.365	-1.4406	-2.3705	-1.5856	-0.4848	-0.6546
4	-4.9979	0.0819	-3.1677	-0.4627	0.911	-0.909	0.5341	-0.2885
5	3.375	-0.4348	0.3218	0.6203	0.875	0.2899	-2.1628	0.5289
6	-2.079	0.9224	1.1297	-1.4267	-2.559	0.0485	0.3766	1.4228
7	1.9947	-0.728	1.0152	-1.3862	-2.13	-5.4382	-2.115	-0.7537
8	-2.8577	0.9839	-1.2675	-2.0439	-1.2717	-0.4118	-3.8042	3.1307

(a)

	65	66	67	68	69	70	71	72
65	841	302.1462	104.7651	30.5575	14	1.2765	2.0653	-3.71
66	456.0342	-108.9023	-130.2682	-72.406	-26.4795	-12.297	-1.2748	-2.4157
67	66.1055	-185.9688	4.5585	70.4124	50.5773	25.3067	9.7522	3.9413
68	-9.0434	1.4466	82.8067	14.0929	-40.3098	-36.1525	-22.282	-5.1615
69	-53.5	36.4496	0.9816	-48.9929	-18.5	25.8295	27.1944	17.2469
70	-18.8813	43.8048	-19.3503	-6.3284	22.7722	15.804	-7.3555	-15.91
71	-4.3809	6.0429	-24.2478	16.6677	9.0866	-17.6497	-12.0585	-7.9933
72	8.8122	2.4084	-10.0236	16.1536	-4.5833	-12.5494	17.4176	20.0054
73	127.625	14.7686	3.4461	5.8781	4.875	3.6376	2.7668	0.8894

(b)

	129	130	131	132	133	134	135	136
129	873.375	-697.752	105.665	94.4273	-58.125	55.9427	-9.8078	-69.514
130	-9.4687	12.7594	13.1746	-6.7241	-12.3344	9.65	-1.1011	-10.2776
131	-8.2965	-1.3019	-3.4205	3.2375	3.1591	8.2861	4.3865	7.6099
132	-0.8682	-3.3157	-5.7457	4.4177	5.0785	2.4324	4.3479	5.5348
133	-2.125	3.1865	1.2225	-2.0555	-2.125	-0.7872	-0.6417	-1.3925
134	3.5626	2.3313	1.4306	-0.8428	-2.8238	-3.0746	-3.3014	-1.778
135	-0.7577	2.7431	0.3865	0.1223	2.0739	-3.4376	-1.8295	3.0859
136	-2.3367	-2.0527	-0.6625	-0.6876	1.2878	2.9097	1.2413	-0.1025

(c)

	249	250	251	252	253	254	255	256
249	917.375	-20.4205	-35.3913	0.9938	-20.875	22.5457	-10.6414	-14.7104
250	-17.4023	41.9748	13.6701	4.1349	8.2	-3.1763	-20.444	-16.5316
251	52.8833	22.1493	-26.8932	-1.3844	-4.0842	16.7127	12.1564	10.6299
252	52.9053	-19.3133	-22.4387	26.6849	9.1569	9.4899	-3.4021	13.551
253	9.625	-3.1002	31.3595	24.3583	-1.125	16.7293	7.4406	14.5476
254	-14.8952	-3.4955	13.5986	3.7178	3.9666	19.7072	1.8681	10.6729
255	-13.1105	-11.8632	-18.3436	-10.7882	17.2511	15.6234	-5.8568	-12.5538
256	-4.9628	14.6963	7.898	-18.8098	2.6907	18.0795	17.4838	-23.8669

(d)

FIGURE 8.5 8 × 8 matrix blocks from Figure 8.4 with four bottom rows and four right columns forced to zero values.

at all. Hence, the actual image size is now reduced to half the size, that is, from 256 × 256 to 128 × 128.

The resulting image is shown in Figure 8.6(a), and its inverse DCT is shown in Figure 8.6(b). It can be seen that the effect of this data loss

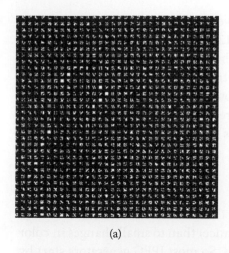

(a)   (b)

FIGURE 8.6 DCT blocks of Figure 8.3(a) with half of the blocks forced to zero, (b) resulting image from inverse DCT on each of these blocks.

appears as some blurred edges and loss of details. This is expected from a lossy compression methodology such as the one presented here.

8.6 ENTROPY CODING

After the data has been quantized into a finite set of values, it can be encoded using an entropy coder to give additional compression. By entropy, we mean the amount of information present in the data, and an entropy coder encodes the given set of symbols with the minimum number of bits required to represent them. Two of the most popular entropy coding schemes are Huffman coding and arithmetic coding. These techniques allocate variable-length codes for specific symbols (coefficient values in our case) in a set of values to be encoded. This saves even more space in terms of storage, thus adding to the compression.

8.7 JPEG COMPRESSION

Technically speaking, JPEG is not a file format at all; it is a compression method that is used in file formats such as TIFF and BMP, and the filename extension is .jpg or .jpeg. JPEG (pronounced "jay-peg") gets its name from the committee that designed it, the Joint Photographic Experts Group. The key phrase here is *photographic experts*; this group was formed especially to design a better means of storing photographic and other photorealistic images. True color (24-bit color) was almost a given, so they concentrated on coming up with a good compression methodology. JPEG uses 24-bit color for RGB images. You can also save standard grayscale images (8 bits per pixel), extended grayscale images (12 bits per pixel), and CMYK images for four-color printers (32-bit color). The advantage of JPEG over GIF and most other methods is this: It provides true color while compressing image files more than the lossless compression methods. The disadvantage is, of course, that it must use a lossy compression method to do that, so some image data is lost, but not as much as when you reduce a photorealistic image to 256 colors.

8.7.1 JPEG's Algorithm

The JPEG method is based on the fact that humans are much more aware of small changes in brightness (luminance) than to small changes in color or large changes in color or brightness. So most JPEG generators start by converting the color data from RGB to a system that identifies the brightness of each pixel. One such system is known as HSL (Hue-Saturation-Luminance). A few other systems also can be used with JPEG such as HSV (Hue-Saturation-Value) and YC_bC_r, where Y is the luminance scale and C_b and C_r are color scales. Once the conversion is made, the first data reduction takes place in a process called *subsampling*. The brightness scale is left alone, while half of the other two scales are eliminated by replacing two neighboring pixels with a single value representing their average. This reduces the entire image to two-thirds of its original size with no noticeable loss in quality, because the most important information—the brightness of each pixel—is still intact.

Some programs average the color scales both vertically and horizontally, thus eliminating three-fourths of the values and reducing the entire image to one-half its original size.

The next steps work on 8-by-8 blocks in the image. In a series of DCT and *quantization* steps, the changes in brightness and color within each block are identified and rounded off. The amount of rounding increases with the size of the change, thus giving more weight to the smaller changes. For example, a change of 3 might be rounded up to 5, whereas a change of 75 might be rounded up to 100.

The result of the previous steps is to produce a set of values that still describes the image data, although not as accurately as before, and that contains many identical values because of the rounding process. All those identical values can now be greatly compressed via a standard, legally unencumbered lossless compression technique.

Because JPEG trades accuracy for compression, the more the file is compressed, the more quality is lost. Three factors affect the quality of the JPEG image:

- The amount of data removed during subsampling (one-third or one-half of the file).

- How aggressively the data is rounded during quantization.

- The accuracy of the JPEG viewer in reversing the compression process while expanding the image for display. There are a lot of calculations involved, and some viewers trade accuracy for speed.

8.8 ALGORITHMIC ACCOUNT

The basic DCT and inverse DCT operations can be performed on images in any programming environment. The main logic used in these cases is shown as flow charts in Figures 8.7 and 8.8.

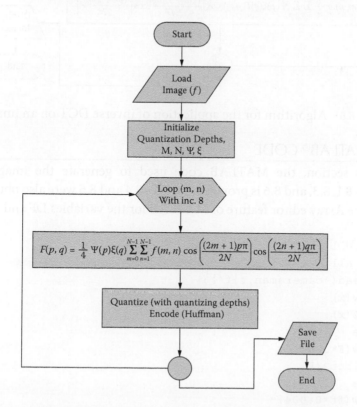

FIGURE 8.7 Algorithm for the application of DCT on an image.

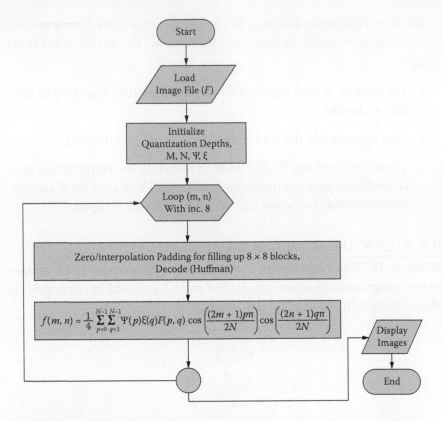

FIGURE 8.8   Algorithm for the application of inverse DCT on an image.

8.9 MATLAB® CODE

In this section, the MATLAB code used to generate the images in Figures 8.1, 8.3, and 8.6 is provided. Figures 8.4 and 8.5 were also obtained from the Array editor feature of MATLAB for the variables DF and IDF.

```
close all
clear all
x=imread('cameraman.tif');
imshow(x)
F=dct2(x);
figure
imshow(F*0.01)
ff = idct2(F) ;
figure
imshow(ff*0.004)
[r,c] = size(x)  ;
```

```
DF = zeros(r,c) ;
DFF = DF ;
IDF = DF ;
IDFF = DF ;
depth = 4 ;
N = 8 ;

for i = 1 : N : r
    for j = 1 : N : c
        f = x(i:i+N-1,j:j+N-1) ;
        df = dct2(f) ;
        DF(i:i+N-1,j:j+N-1) = df ;
        dff = idct2(df) ;
        DFF(i:i+N-1,j:j+N-1) = dff ;
        df(N:-1:depth+1,:) = 0 ;
        df(:,N:-1:depth+1) = 0 ;
        IDF(i:i+N-1,j:j+N-1) = df ;
        dff = idct2(df) ;
        IDFF(i:i+N-1,j:j+N-1) = dff ;
    end
end

figure
imshow(DF)
figure
imshow(DFF*0.005)

figure
imshow(IDF)
figure
imshow(IDFF*0.004)
```

8.10 SUMMARY

- Image compression refers to the technique in which image pixel values are mapped into a data space from where they can be stored using a more compact structure.

- The first level of mapping usually comprises some type of transformation from pixel domain to coefficient domain of some type.

- These coefficients are then quantized in magnitude as well as in terms of their locations.

- Some form of variable-length coding technique is then used to encode the final coefficients.

- For decompression, the reverse procedure is utilized.

- The most commonly used transforms are the discrete wavelet transform and discrete cosine transform.

- The most commonly used coding techniques are Huffman coding and arithmetic coding.

- The JPEG compression standard utilizes all of the aforementioned techniques.

- If information is heavily lost in the quantization step, the decompression will not produce the original image, and the compression is termed *lossy compression*.

8.11 EXERCISES

1. Repeat the exercise done on the cameraman image using blocks of size 16 and 32 with depths of 8 and 16, respectively. This will essentially give the same compression ratio. Comment on the differences from the block size 8 method in terms of quality of decompressed image and time of calculation.

2. Use an RGB image and apply the DCT-based compression routine as given in Section 8.9 on each layer; note down the compression as well as the quality difference for the decompressed image.

3. Convert the image in Q #2 to grayscale first, and then apply DCT-based compression and decompression. Compare the results of Q #2 and Q #3 in terms of whether RGB-based compression is less efficient than grayscale-based compression.

4. Convert the image in Q #2 to YC_bC_r format. Apply block size 16 with DCT compression on the Y component with depth of 4, and for the remaining component, apply a depth of 12. Then decompress it to a normal color image and compare the color quality of the image before and after the application of the algorithm.

5. Can the procedure in Q #4 be applied to an indexed image? If so, which part of the image should be compressed using 8-DCT: the main indexed image or the map?

CHAPTER 9

Edge Detection

9.1 INTRODUCTION

THE BIGGEST GROUP OF processing operators exists for a class of images called the *binary image*. This is especially true for machine vision applications, where most of the color- and texture-related information is not of much use. In these applications, only certain areas of interest are isolated to be processed for further stages of processing. However, conversion to binary images is one of the essential preprocessing steps needed in machine vision processing. As discussed in Chapter 2, Section 2.5, a grayscale image can be converted to a binary image by thresholding the grayscale levels into two levels only: black and white. The next important information to be extracted from a grayscale image is related to the edges present in the image.

An edge can be defined as "a sudden change of intensity in an image." In binary images, an edge corresponds to a sudden change in intensity level to 1 from 0, and vice versa. This, essentially, represents high-frequency components in the image and, thus, extracting them would involve some procedure to extract high frequencies from the image. Hence, edge detection is fundamentally a high-pass operation with thresholding added to it. This implies that once high-frequency components (i.e., edges) are found, they are thresholded to logic 1 in the image while all other pixels are set to zero. Too many edges represent a "rough" image, while too few would correspond to a relatively "smoother" image. The following sections discuss some of the common techniques used for edge detection.

−1	0	+1
−2	0	+2
−1	0	+1

G_x

+1	+2	+1
0	0	0
−1	−2	−1

G_y

f_1	f_2	f_3
f_4	f_5	f_6
f_7	f_8	f_9

FIGURE 9.1   Sobel convolution kernels, G_x (left) and G_y (middle), and image segment (right) to be operated on with Equation 9.1.

9.2 THE SOBEL OPERATOR

The Sobel operator performs a 2D spatial gradient measurement on an image and emphasizes regions of high spatial frequency that correspond to edges. Typically, it is used to find the approximate absolute gradient magnitude at each point in an input grayscale image. The operator consists of a pair of 3 × 3 convolution kernels, as shown in Figure 9.1. One kernel is simply the other rotated by 90°. Assuming an image segment of the same dimensions as that of the Sobel kernel (as shown in Figure 9.1), the result of convolution will be

$$|G|=|(f_1+2f_2+f_3)-(f_7+2f_8+f_9)|+|(f_1+2f_4+f_7)-(f_3+2f_6+f_9)| \quad (9.1)$$

These kernels are designed to respond maximally to edges running vertically and horizontally relative to the pixel grid, one kernel for each of the two perpendicular orientations. The kernels can be applied separately to the input image to produce separate measurements of the gradient component in each orientation (call these G_x and G_y). These can then be combined together to find the absolute magnitude of the gradient at each point and the orientation of that gradient. The gradient magnitude is given by

$$|G|= \sqrt{G_x^2+G_y^2}. \quad (9.2)$$

Typically, an approximate magnitude is computed for faster calculations using

$$|G|=|G_x|+|G_y|. \quad (9.3)$$

The angle of orientation of the edge (relative to the pixel grid) giving rise to the spatial gradient is given by

$$\theta = \tan^{-1}\left(\frac{G_y}{G_x}\right). \tag{9.4}$$

In this case, orientation 0 is taken to mean that the direction of maximum contrast from black to white runs from left to right on the image, and

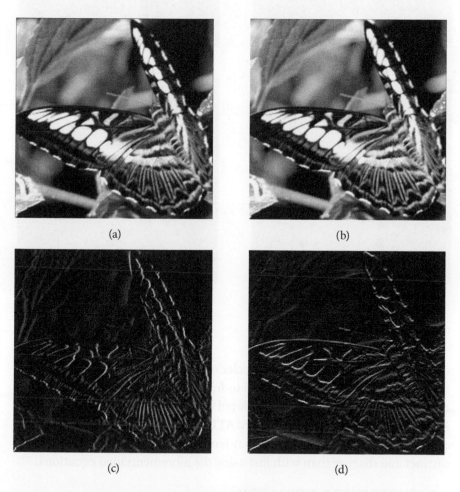

(a)

(b)

(c)

(d)

FIGURE 9.2 Application of the Sobel operator. (a) Original RGB image, (b) grayscale image, (c) G_x, (d) G_y, (e) image using Equation 9.2, (f) image using Equation 9.3, (g) MATLAB's Sobel operator with threshold 0.1, (h) MATLAB's Sobel operator with threshold 0.05.

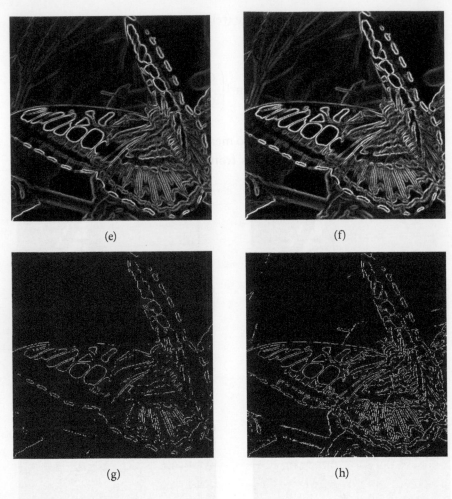

(e)   (f)

(g)   (h)

FIGURE 9.2 (Continued)

other angles are measured counterclockwise from this. Figure 9.2 shows how MATLAB® functions are used to implement the Sobel operators. The components G_x and G_y are also displayed separately, as well as the results for Equations 9.2 and 9.3. Usually, the MATLAB functions perform some type of morphological postprocessing that renders the edges in the image more distinct and sharper than with just using the aforementioned equations.

9.3 THE PREWITT OPERATOR

This works in a very similar way to the Sobel operator but uses slightly different kernels, as shown in Figure 9.3. This kernel produces similar results to the Sobel, but is not as isotropic in its response.

−1	0	+1
−1	0	+1
−1	0	+1

G_x

+1	+1	+1
0	0	0
−1	−1	−1

G_y

FIGURE 9.3   Prewitt operator kernels.

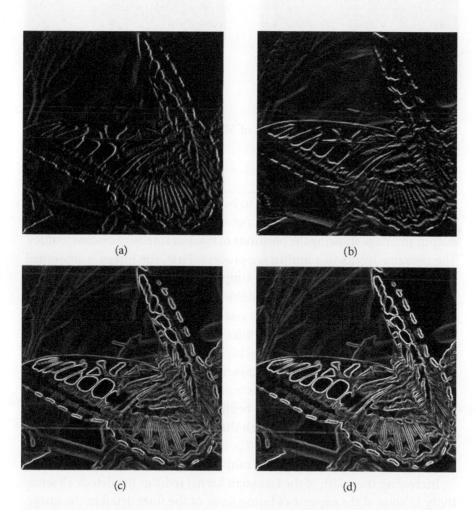

(a)

(b)

(c)

(d)

FIGURE 9.4   Application of Prewitt operator. (a) G_x, (b) G_y, (c) image using Equation 9.2, (d) image using Equation 9.3, (e) MATLAB's Prewitt operator with threshold 0.1, (f) MATLAB's Prewitt operator with threshold 0.05.

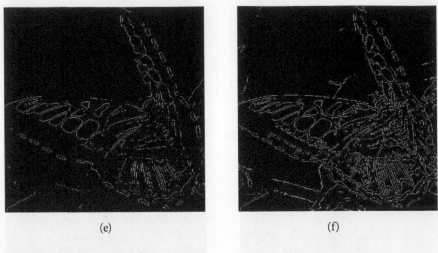

(e)   (f)

FIGURE 9.4 (Continued)

Figure 9.4 shows the application of MATLAB functions for implementing the Prewitt operators.

9.4 THE CANNY OPERATOR

The Canny operator was designed to be an optimal edge detector satisfying particular criteria. It takes as input a grayscale image, and produces as output an image showing the positions of tracked intensity discontinuities. The operator works in a multistage process. First, the image is smoothed by Gaussian convolution. Then a simple 2D first derivative operator is applied to the smoothed image to highlight regions of the image with high first-spatial derivatives. Edges give rise to ridges in the gradient magnitude image. The algorithm then tracks along the top of these ridges and sets to zero all pixels that are not actually on the ridge top so as to give a thin line in the output, a process known as *nonmaximal suppression*. The tracking process exhibits hysteresis controlled by two thresholds: T_1 and T_2, with $T_1 > T_2$. Tracking can only begin at a point on a ridge higher than T_1. Tracking then continues in both directions out from that point until the height of the ridge falls below T_2. This hysteresis helps to ensure that noisy edges are not broken up into multiple edge fragments.

Increasing the width of the Gaussian kernel reduces the detector's sensitivity to noise at the expense of losing some of the finer detail in the image. The localization error in the detected edges also increases slightly as the Gaussian width is increased. Usually, the upper tracking threshold can be set quite high, and the lower threshold quite low for good results. Setting

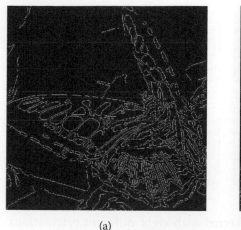
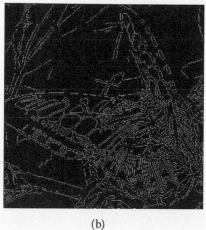

(a)   (b)

FIGURE 9.5   Edge detection results with Canny operator. (a) With a threshold of 0.2, (b) with a threshold of 0.1.

the lower threshold too high will cause noisy edges to break up. Setting the upper threshold too low increases the number of spurious and undesirable edge fragments appearing in the output. Figure 9.5 shows the application of the Canny operator. As can be seen, the edges obtained by the Canny operator are much more smooth and clear and hence more tolerant to noise. On the other hand, it takes a lot longer to calculate Canny edges than the usual convolutional edges. For the images shown, the Canny operator takes 3.3 times longer to calculate the edges than the Sobel operator.

9.5 THE COMPASS OPERATOR (EDGE TEMPLATE MATCHING)

This operator usually outputs two images, one estimating the local edge gradient magnitude and the other estimating the edge orientation of the input image. The source image is convolved with a set of convolution kernels, each of which is sensitive to edges in a different orientation. For each pixel, the local edge gradient magnitude is estimated with the maximum response of all of these kernels at this pixel location:

$$|G| = \max(|G_i|; \qquad i = 1 \quad to \quad n), \qquad (9.5)$$

where G_i is the response of the kernel i at the particular pixel position and n is the total number of convolution kernels. The local edge *orientation* is estimated with the orientation of the kernel that yields the maximum response.

−1	0	+1
−2	0	+2
−1	0	+1

0°

0	+1	+2
−1	0	+1
−2	−1	0

45°

+1	+2	+1
0	0	0
−1	−2	−1

90°

+2	+1	+0
+1	0	−1
0	−1	−2

135°

+1	0	−1
+2	0	−2
+1	0	−1

180°

0	−1	−2
+1	0	−1
+2	+1	0

225°

−1	−2	−1
0	0	0
+1	+2	+1

270°

−2	−1	0
−1	0	+1
0	−1	+2

315°

FIGURE 9.6 Sobel edge detection kernel with eight different orientations to be used with the compass edge detector.

Various kernels can be used for this operation. The whole set of *n* kernels is produced by taking one of the main standard kernels and rotating its coefficients circularly. Each of the resulting kernels is sensitive to an edge orientation ranging from 0° to 360° − Δ in steps of Δ, where 0° corresponds to a vertical edge, and so on.

The compass edge detector is an appropriate way to estimate both the magnitude *and* orientation of an edge. Although differential gradient edge detection needs a rather time-consuming calculation to estimate the orientation from the magnitudes in the *x*- and *y*-directions, the compass edge detection obtains the orientation directly from the kernel with the maximum response. On the other hand, the compass operator needs *n* convolutions for each pixel, whereas the gradient operator needs only 2. As an example, the Sobel operator has been shown in Figure 9.6 with eight different orientations to cover various directional edges, while Figure 9.7 shows some of the commonly used kernels.

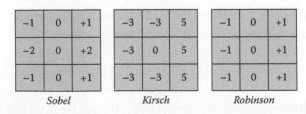

Sobel   Kirsch   Robinson

FIGURE 9.7 Some examples of the most common compass edge-detecting kernels.

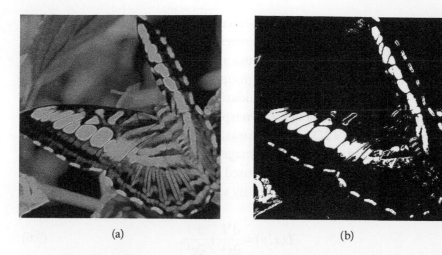

(a)   (b)

FIGURE 9.8 Application of edge template matching. (a) Maximal image, (b) thresholded edges.

Figure 9.8 shows the application of this operator to the image of Figure 9.1(b).

9.6 THE ZERO-CROSSING DETECTOR

The zero-crossing detector looks for places in the Laplacian of an image where the value of the Laplacian passes through zero, that is, points where the Laplacian changes sign. Such points often occur at edges in images. It is best to think of the zero-crossing detector as some sort of feature detector rather than as a specific edge detector.

The starting point for the zero-crossing detector is an image that has been filtered using the Laplacian of Gaussian (LoG) filter. The zero crossings that result are strongly influenced by the size of the Gaussian used for the smoothing stage of this operator. As the smoothing is increased, fewer and fewer zero-crossing contours will be found, and those that do remain will correspond to features of larger and larger scale in the image.

However, zero crossings also occur at any place where the image intensity gradient starts increasing or decreasing, and this may happen at places that are not obviously edges. Often, zero crossings are found in regions of very low gradient where the intensity gradient wobbles up and down around zero.

Once the image has been LoG-filtered, it only remains to detect the zero crossings. This can be done in several ways. The simplest is to threshold the LoG output at zero, to produce a binary image where the boundaries between foreground and background regions represent the locations of zero-crossing points. These boundaries can then be easily detected and marked in single pass using some morphological operator.

The Laplacian is a 2D isotropic measure of the second spatial derivative of an image. The Laplacian of an image highlights regions of rapid intensity change and is therefore often used for edge detection. The Laplacian $L(x,y)$ of an image with pixel intensity values $I(x,y)$ is given by

$$L(x,y) = \frac{\partial^2 I}{\partial x^2} + \frac{\partial^2 I}{\partial y^2}. \tag{9.6}$$

This can be calculated using a convolution filter.

The 2D LoG function centered on zero and with Gaussian standard deviation σ has the form:

$$LoG(x,y) = -\frac{1}{\pi\sigma^4}\left[1 - \frac{x^2+y^2}{2\sigma^2}\right]e^{-\frac{x^2+y^2}{2\sigma^2}}. \tag{9.7}$$

The true shape of the LoG kernel is shown in Figure 9.9.

In fact, because the convolution operation is associative, we can convolve the Gaussian smoothing filter with the Laplacian filter first, and then convolve this hybrid filter with the image to achieve the required result. Doing things this way has two advantages:

- As both the Gaussian and Laplacian kernels are usually much smaller than the image, this method usually requires far fewer arithmetic operations.

- The LoG kernel can be calculated in advance, so only one convolution needs to be performed at run time on the image.

The LoG operator calculates the second spatial derivative of an image. This means that in areas where the image has a constant intensity (i.e., where the intensity gradient is zero), the LoG response will be zero. In the vicinity of a change in intensity, however, the LoG response will be positive on the darker side, and negative on the lighter side. This means that

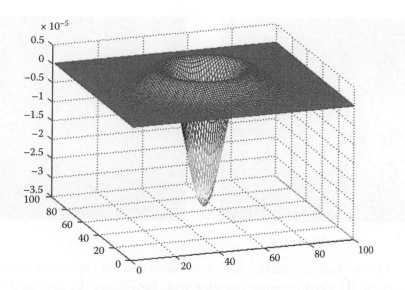

FIGURE 9.9 The 3D realization of the LoG operator.

at a reasonably sharp edge between two regions of uniform but different intensities, the LoG response will be as follows:

- Zero at a long distance from the edge

- Positive just to one side of the edge

- Negative just to the other side of the edge

- Zero at some point in between, on the edge itself

The problem with this technique is that it will tend to bias the location of the zero-crossing edge to either the light side of the edge, or the dark side of the edge, depending on whether it is decided to look for the edges of foreground regions or for the edges of background regions.

A better technique is to consider points on both sides of the threshold boundary, and choose the one with the lowest absolute magnitude of the Laplacian, which will hopefully be closest to the zero crossing.

Because the zero crossings generally fall between two pixels in the LoG-filtered image, an alternative output representation is an image grid that is spatially shifted half a pixel across and half a pixel down, relative to the original image. Such a representation is known as a *dual lattice*. This does not actually localize the zero crossing any more accurately, of course.

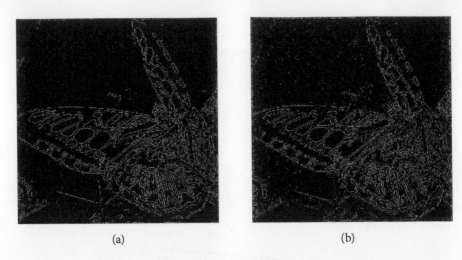

(a)   (b)

FIGURE 9.10 Implementation of zero-crossing and LoG operators for edge detection. (a) Zero-crossing operator with LoG filter, (b) LoG operator alone as applied to the test image.

A more accurate approach is to perform some kind of interpolation to estimate the position of the zero crossing to subpixel precision.

All edges detected by the zero-crossing detector are in the form of closed curves in the same way that contour lines on a map are always closed. The only exception to this is where the curve goes off the edge of the image.

Because the LoG filter is calculating a second derivative of the image, it is quite susceptible to noise, particularly if the standard deviation of the smoothing Gaussian is small. Thus, it is common to see lots of spurious edges detected away from any obvious edges. One solution to this is to increase the smoothing of the Gaussian to preserve only strong edges. Another is to look at the gradient of the LoG at the zero crossing (i.e., the third derivative of the original image) and only keep zero crossings where this is above a certain threshold. This will tend to retain only the stronger edges, but it is sensitive to noise, because the third derivative will greatly amplify any high-frequency noise in the image. Figure 9.10 shows the implementation of zero-crossing and LoG operators for edge detection.

9.7 LINE DETECTION

Although edges are by far the most common type of discontinuity in an image, instances of thin lines in an image occur frequently enough that it is useful to have a separate mechanism for detecting them. Note that the

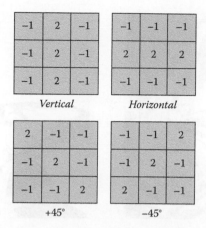

FIGURE 9.11   Four line detection kernels that respond maximally to horizontal, vertical, and oblique (+45 and −45°) single-pixel-wide lines.

Hough transform can also be used to detect lines; however, in that case, the output is a *parametric* description of the lines in an image that would need more processing to be finally converted to the lines. Similar to the edge detection operator, the line detection operator consists of a convolution kernel tuned to detect the presence of lines of a particular width n, at a particular orientation θ. Figure 9.11 shows a collection of four such kernels, which respond to lines of single pixel width at the particular orientation shown.

These masks are tuned for light lines against a dark background, and would give a big negative response to dark lines against a light background. If one is only interested in detecting dark lines against a light background, then one should negate the mask values or the source image before applying these kernels.

9.8 THE UNSHARP FILTER

The unsharp filter is a simple sharpening operator that derives its name from the fact that it enhances edges (and other high-frequency components in an image) via a procedure that subtracts an unsharp, or smoothed, version of an image from the original image. The unsharp filtering technique is commonly used in the photographic and printing industries for crispening edges.

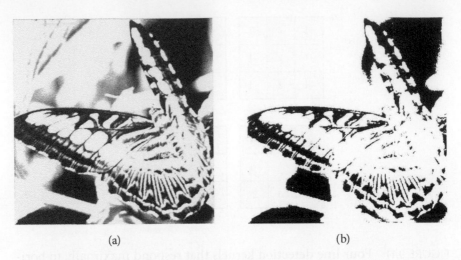

(a)   (b)

FIGURE 9.12 Application of unsharp filtering operator. (a) Test image, (b) the binary version of (a).

Unsharp masking produces an edge image $g(m,n)$ from an input image $f(m,n)$ via

$$g(m,n) = f(m,n) - f_{\text{smooth}}(m,n), \qquad (9.8)$$

where $f_{\text{smooth}}(m,n)$ is a smoothed version of $f(m,n)$, which results after applying a simple low-pass filter to the image. Figure 9.12 shows the application of the unsharp filtering operator on the test image.

9.9 ALGORITHMIC ACCOUNT

As can be observed from the foregoing discussion, three main approaches have been commonly utilized in the image processing world for edge detection. These are shown as logic flow charts in Figure 9.13.

In general, the first category is related to the convolution kernels (such as Sobel), in which a specific kernel is initialized in the beginning and is then convolved with the image, enhancing the edges, which are then converted to binary through a thresholding step. The second category is a combination of various convolution kernels and their application in the usual way of (i.e., same as the first category). However, for each kernel, the results are compared and stored for the maximal response. An example of such a technique is the compass edge detector. In the third category, an entirely different approach of gradient calculation is utilized. First, a smoothing step is performed, followed by a gradient calculation; then the

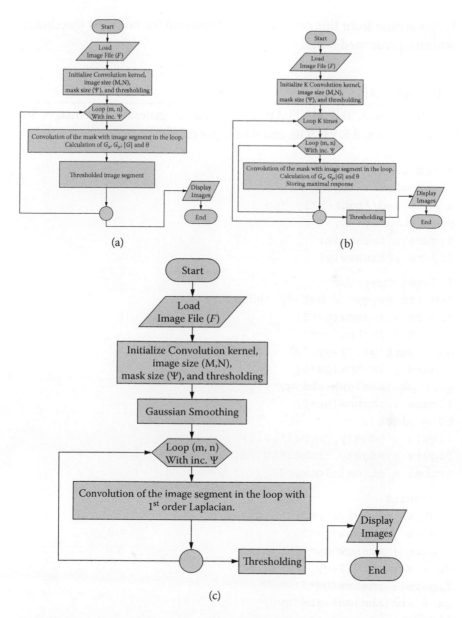

FIGURE 9.13 Algorithms for edge detection using (a) convolutional kernels, (b) compass edge detector, (c) the Canny operator.

ridges arising from this calculation are traversed for bounded thresholds within a predefined range.

9.10 MATLAB CODE

In this section, the MATLAB code used to generate the images in Figures 9.2, 9.4, 9.5, 9.9, 9.10, and 9.12 is provided.

```
close all
clear all
x = imread('butterfly.jpg');
y = rgb2gray(x);
figure ; imshow(x)
figure ; imshow(y)

% Sobel Operator
[ed1,th,gx,gy] = edge(y,'sobel');
figure ; imshow(gx*5)
figure ; imshow(gy*5)
gs = sqrt(gx.^2+gy.^2) ;
figure ; imshow(gs*5)
ga = abs(abs(gx)+abs(gy)) ;
figure ; imshow(ga*5)
t0 = clock ;
figure ; edge(y,'sobel',0.1) ;
figure ; edge(y,'sobel',0.05) ;
tsobel = etime(clock,t0) ;

% Prewitt
[ed1,th,gx,gy] = edge(y,'prewitt');
figure ; imshow(gx*5)
figure ; imshow(gy*5)
gs = sqrt(gx.^2+gy.^2) ;
figure ; imshow(gs*5)
ga = abs(abs(gx)+abs(gy)) ;
figure ; imshow(ga*5)
figure ; edge(y,'prewitt',0.1) ;
figure ; edge(y,'prewitt',0.05) ;

t0 = clock ;
figure ; edge(y,'canny',0.1) ;
figure ; edge(y,'canny',0.05) ;
tcanny = etime(clock,t0) ;
```

```
% Compass edge
sx1 = [-1 0 1 ; -2 0 2 ; -1 0 1] ;
sa1 = [ 0 1 2 ; -1 0 1 ; -2 -1 0 ] ;
sx2 = rot90(sx1) ; sa2 = rot90(sa1) ;
sx3 = rot90(sx2) ; sa3 = rot90(sa2) ;
sx4 = rot90(sx3) ; sa4 = rot90(sa3) ;
sy1 = [1 2 1 ; 0 0 0 ; -1 -2 1] ;
sb1 = [ 2 1 0 ; 1 0 -1 ; 0 -1 -2 ] ;
sy2 = rot90(sy1) ; sb2 = rot90(sb1) ;
sy3 = rot90(sy2) ; sb3 = rot90(sb2) ;
sy4 = rot90(sy3) ; sb4 = rot90(sb3) ;

g1x = conv2(y,sx1,'same') ; g1y = conv2(y,sy1,'same') ;
g2x = conv2(y,sa1,'same') ; g2y = conv2(y,sb1,'same') ;
g3x = conv2(y,sx2,'same') ; g3y = conv2(y,sy2,'same') ;
g4x = conv2(y,sa2,'same') ; g4y = conv2(y,sb2,'same') ;
g5x = conv2(y,sx3,'same') ; g5y = conv2(y,sy3,'same') ;
g6x = conv2(y,sa3,'same') ; g6y = conv2(y,sb3,'same') ;
g7x = conv2(y,sx4,'same') ; g7y = conv2(y,sy4,'same') ;
g8x = conv2(y,sa4,'same') ; g8y = conv2(y,sb4,'same') ;

g1 = sqrt(g1x.^2 + g1y.^2) ;
g2 = sqrt(g2x.^2 + g2y.^2) ;
mg = max(g1,g2) ;
g3 = sqrt(g3x.^2 + g3y.^2) ;
mg = max(mg,g3) ;
g4 = sqrt(g4x.^2 + g4y.^2) ;
mg = max(mg,g4) ;
g5 = sqrt(g5x.^2 + g5y.^2) ;
mg = max(mg,g5) ;
g6 = sqrt(g6x.^2 + g6y.^2) ;
mg = max(mg,g6) ;
g7 = sqrt(g7x.^2 + g7y.^2) ;
mg = max(mg,g7) ;
g8 = sqrt(g8x.^2 + g8y.^2) ;
mg = max(mg,g8) ;
M = mg*0.0005 ;
figure ; imshow(M)
MM = im2bw(M,0.25) ;
figure ; imshow(MM)

% LoG and Zero-Crossing
h = fspecial('log',[100 100],10) ;
```

```
figure ; mesh(h) ; colormap hsv
h = fspecial('log',[10 10],1.5) ;
figure ; edge(y,'zerocross',h)
figure ; edge(y,'log',0.01,1)
% Unsharp
h=fspecial('unsharp',0.2);
G = conv2(y,h,'same')*0.01 ;
figure ; imshow(G)
figure ; im2bw(G,0.2)
```

9.11 SUMMARY

- Edges in an image are defined as the sudden changes in the intensity levels from dark to light, and vice versa.

- Edge detection is a key operation in many machine vision and automation systems because this information could extract the objects of interest as well as regions of particular importance from an image.

- Being spatially high-frequency components, the edge detection strategies usually correspond to high-pass filtering of the image to allow these high-frequency components to pass to the next stage.

- Because typical filter implementation in images is through convolution kernels, the edge detectors are also similar in nature.

- Almost all of these kernels calculate two submatrices with horizontal and vertical edges enhanced, respectively. Then, an overall edge plane is calculated for which a binary thresholding is performed to produce the edges.

- Commonly used edge detectors are Sobel, Prewitt, Robert, Robinson, LoG, zero-crossing, and Kirsch operators.

- Because each of the aforementioned operators is designed for a particular direction, a combinational technique is used in the compass edge detector method.

- In this technique, any standard operator is rotated for K directional orientations, and the image is convolved for each of them. The maximum of these convolutions are stored as final edges.

- A better approach is used in Canny operation to first smooth out the image, after which gradients are calculated.

- On these gradient peaks, a calculated thresholding is applied to produce the final edges.

- The Canny operator is very robust to noise compared to the other operators, but is slowest of all.

- Straight lines can also be detected in an image using similar kernels as those of convolution kernels.

- A specialized sharpening filter can also be used first to enhance the edges before performing the thresholding operation.

- The LoG operator calculates the second spatial derivative of an image. This means that in areas where the image has a constant intensity (i.e., where the intensity gradient is zero), the LoG response will be zero.

- In the vicinity of a change in intensity, however, the LoG response will be positive on the darker side, and negative on the lighter side.

9.12 EXERCISES

1. Some statistical features of edge detection can be used for other applications in machine vision systems. For instance, in an iterative filtering or enhancement procedure, the image is improved in every iteration until it starts to deteriorate again. Such features can be used to detect this point and, hence, can be used as a stoppage criterion. Use a noisy image, and iteratively apply Gaussian filtering on it. After each filtering, calculate the edges of the image using the Canny operator. For the resulting binary image, calculate the following:

   a. Entropy of the binary image (all edges): $E = \Sigma p_i \log_{10} p_i$, where p_i is the probability of ith intensity in the image (equivalent to the ith bin in the image's histogram)

   b. Assign the threshold of 0.5 (fixed).

2. Repeat Q #1 by assigning the threshold (T) a linear iterative weight: $(i - 1)T/10$, where i is the iteration index starting from 1.

3. Repeat Q #1 by assigning the threshold (T) an exponential iterative weight $e^{-(i-1)}T$, where i is the iteration index starting from 1.

4. In Q #1, what other statistical measure that can be used instead of entropy?

5. Read an RGB image and apply the Sobel operator on each layer separately. Comment on the edge localization in terms of their similarity/dissimilarity for each layer. Also, display the new RGB image by keeping the new three layers as one image entity. Comment on the new color scheme.

CHAPTER 10

Binary Image Processing

10.1 INTRODUCTION

OTHER THAN EDGE DETECTION, the main processing performed on purely binary images can be grouped into a broader category called *morphological operators*. They process objects in the input image based on characteristics of its shape, which are encoded in the structuring element. Usually, each morphological operator would work on a 3 × 3 neighborhood in which it applies the laws of the operator under consideration. It is then shifted over the image, and at each pixel of the image, its elements are compared with the set of the underlying pixels. If the two sets of elements match the condition defined by the set operator, the pixel under the origin of the neighborhood (the center pixel) is set to a predefined value (0 or 1 for binary images). Morphological operators can also be applied to gray-level images, for example, to reduce noise or to brighten the image. However, for many applications, other methods, such as a more general spatial filter, produce better results.

In this chapter, the following morphological operators are discussed:

1. Dilation and erosion

2. Closing and opening

3. Thickening and thinning

4. Skeletonization

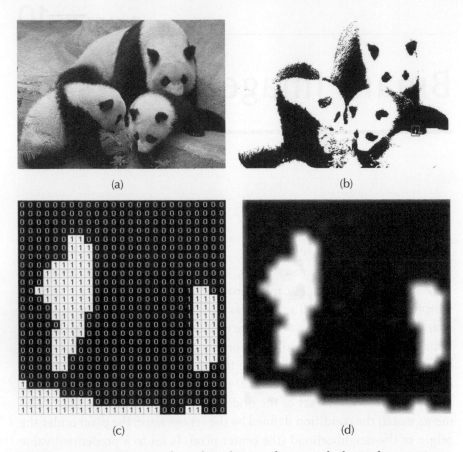

FIGURE 10.1 Images used in this chapter for morphological operations. (a) Original RGB image, (b) binary version of (a) with 50% threshold, (c) actual values for the selected segment, (d) selected segment of the image.

Each member of a pair in the foregoing list (i.e items 1 through 3) performs opposite functions. Essentially, dilation, closing, and thickening would make the white elements in the binary image more predominant in various degrees, and similarly, erosion, opening, and thinning would reduce the size of these white elements accordingly. To increase the mass of a white element in the image, more 1's are added while for mass reduction, 0s are added. "Adding" these new values means that opposite pixels are replaced by these new values (*no* physical addition is carried out). All of these operators work on binary images, although their slightly modified versions can also be used with grayscale images.

In the following sections, the image shown in Figure 10.1(a) is used with its binary version [Figure 10.1(b)] for the aforementioned morphological operations. For in-depth visualization, a portion of the image (one of the paws that is visible on the right side of the image) is taken out as a 25 × 25 binary image, and its content is displayed separately to explain the changes occurring in the image as a result of morphological operations. The segment of the image is shown in Figure 10.1(d), and its matrix version is shown in Figure 10.1(c).

Another common feature of all the operators is a structuring element (strel). In order to explain the algorithms mathematically, an easier approach would be to consider the same operations (masking, deciding, replacing pixels, shifting) in all morphological operators but change the strel according to the operation. Throughout this chapter, a default strel of size 3 × 3 is used.

10.2 DILATION

The basic effect of the dilation operator on a binary image is to gradually enlarge the boundaries of regions of foreground pixels (i.e., white pixels, typically). Thus, areas of foreground pixels grow in size while holes within those regions become smaller. Assume that X is the set of Euclidean coordinates corresponding to the input binary image, and that K is the set of coordinates for the strel; let K_x denote the translation of K so that its origin is at x. Then, the dilation of X by K is simply the set of all points x such that the intersection of K_x with X is nonempty. The strel is a 3 × 3 matrix of all ones for the dilation operation. The mathematical definition of grayscale dilation is identical except for the way in which the set of coordinates associated with the input image is derived. In addition, these coordinates are 3D rather than 2D.

To compute the dilation of a binary input image by this structuring element, we consider each of the *background* pixels in the input image in turn. For each background pixel (which we will call the *input pixel*), we superimpose the structuring element on top of the input image so that the origin of the structuring element coincides with the input pixel position. If *at least one* pixel in the structuring element coincides with a foreground pixel in the image underneath, the input pixel is set to the foreground value. If all the corresponding pixels in the image are background, however, the input pixel is left at the background value.

Figure 10.2 shows the effect of application of the dilation operation. The image in Figure 10.2(a) shows the sample image from Figure 10.1(b) with the dilation operator applied thrice, while the image segment shows the effect after a single application of the operator. This anomaly was decided upon experimentally because one iteration of dilation

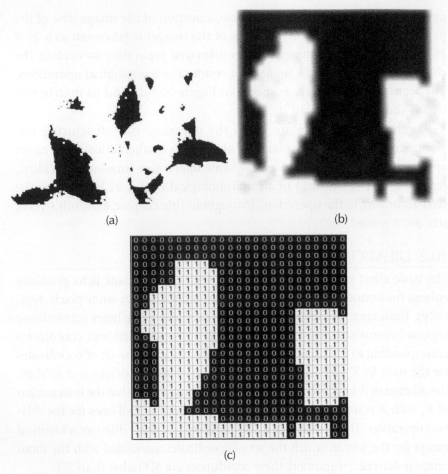

FIGURE 10.2 Application of the dilation operator. (a) Resulting image after three iterations, (b) resulting image segment after one iteration, (c) actual values for (b).

may not appear significantly visible in a large image [such as the one in Figure 10.1(b)]. On the other hand, the smaller image is used to demonstrate the mechanism of the operation and how it changes the pixels in the original image in one iteration. This convention will be followed throughout this chapter.

Grayscale images can also be used with the dilation operation by using a flat, disk-shaped strel as opposed to an all-1s square and will generally brighten the image. Bright regions surrounded by dark regions grow in size, and dark regions surrounded by bright regions shrink in size. Small, dark spots in images will disappear as they are "filled in" to the surrounding intensity value. Small, bright spots will become larger spots. The effect is most marked at places in the image where the intensity changes rapidly, and regions of fairly uniform intensity will be largely unchanged except at their edges.

10.3 EROSION

Erosion is the opposite of dilation. The basic effect of the operator on a binary image is to erode away the boundaries of regions of foreground pixels (i.e., white pixels, typically). Thus, areas of foreground pixels shrink in size, and holes within those areas become larger. The effect of this operation is to remove any foreground pixel that is not completely surrounded by other white pixels (assuming 8-connectedness). Such pixels must lie at the edges of white regions, and so the practical upshot is that foreground regions shrink (and holes inside a region grow). The strel for the erosion operation is also a 3×3 matrix of 1s, but the action taken is the reverse of the one for dilation. The effect of application of the erosion operator is shown in Figure 10.3.

10.4 OPENING

The basic effect of an opening is somewhat similar to erosion in that it tends to remove some of the foreground (bright) pixels from the edges of regions of foreground pixels. However, in general, it is less destructive than erosion. An opening is simply defined as an erosion followed by a dilation *using the same strel for both operations*. The opening operator therefore requires two inputs: an image to be opened and a strel.

Although erosion can be used to eliminate small clumps of undesirable foreground pixels, for example, speckle noise, quite effectively, it has the big disadvantage that it will affect *all* regions of foreground pixels indiscriminately. Opening gets around this by performing both an erosion and a dilation on the image. The effect of opening can be quite easily visualized. Imagine taking the strel and sliding it around *inside* each foreground region, without changing its orientation. All pixels that can be covered by the structuring element with the structuring element being entirely within the foreground region will be preserved. However, all foreground pixels that cannot be reached by the strel without parts of it moving out of the foreground region will be eroded away. After the opening operation has been carried out, the new boundaries of foreground regions will all be such that the strel fits inside them, and so further

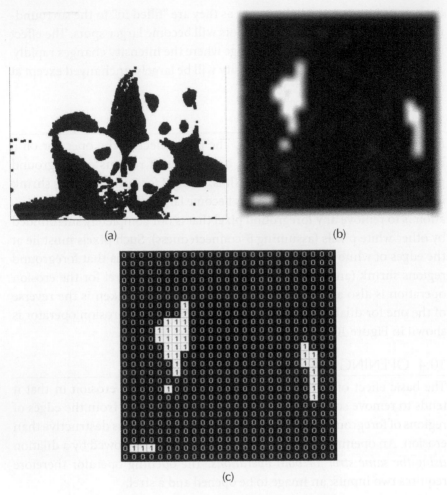

FIGURE 10.3 Application of the erosion operator. (a) Resulting image after three iterations, (b) resulting image segment after one iteration, (c) actual values for (b).

openings with the same element have no effect. Figure 10.4 shows the application of this operation.

10.5 CLOSING

Like its dual operator, opening, the closing operation can be derived from the fundamental operations of erosion and dilation. Closing is similar in some ways to dilation in that it tends to enlarge the boundaries of foreground (bright) regions in an image (and shrink background color holes

FIGURE 10.4 Application of the opening operator. (a) Resulting image after three iterations, (b) resulting image segment after one iteration, (c) actual values for (b).

in such regions), but it is less destructive of the original boundary shape. Closing is the opening operation performed in reverse. It is defined simply as a dilation followed by an erosion using the same strel for both operations. One of the uses of dilation is to fill in small background color holes in images. One of the problems with doing this, however, is that the dilation will also distort *all* regions of pixels indiscriminately. By performing an erosion on the image after the dilation, that is, a closing, we reduce

FIGURE 10.5 Application of the closing operator. (a) Resulting image after three iterations, (b) resulting image segment after one iteration, (c) actual values for (b).

some of this effect. Figure 10.5 shows the result of application of this operator on the sample image.

10.6 THINNING

Thinning is a morphological operation that is used to remove selected foreground pixels from binary images, somewhat like erosion or opening. It can be used for several applications, but is particularly useful for skeletonization. In this mode, it is commonly used to tidy up the output of edge detectors by reducing all lines to single-pixel thickness. Thinning is normally only applied to binary images, and produces another binary image as output.

The thinning operation is calculated by translating the origin of the structuring element to each possible pixel position in the image, and at each such position, comparing it with the underlying image pixels. If the foreground and background pixels in the structuring element exactly match foreground and background pixels in the image, then the image pixel underneath the origin of the structuring element is set to background (zero). Otherwise, it is left unchanged. Note that the strel must always have a one or a blank at its origin if it is to have any effect.

The choice of strel determines under what situations a foreground pixel will be set to background and, hence, it determines the application for the thinning operation.

We have described the effects of a single pass of a thinning operation over the image. In fact, the operator is normally applied repeatedly until it causes no further changes to the image (i.e., until *convergence*). Alternatively, in some applications, for example, *pruning*, the operations may only be applied for a limited number of iterations.

Consider all pixels on the boundaries of foreground regions (i.e., foreground points that have at least one background neighbor). Delete any such point that has more than one foreground neighbor, as long as doing so does not *locally disconnect* (i.e., split into two) the region containing that pixel. Iterate until convergence. This procedure erodes away the boundaries of foreground objects as much as possible, but does not affect pixels at the ends of lines. Figure 10.6 shows some of the commonly used strels for thinning.

Figure 10.7 shows the application of the thinning operator to the sample image.

10.7 THICKENING

Thickening is a morphological operation that is used to *grow* selected regions of foreground pixels in binary images, somewhat similar to dilation or closing. It has several applications, including determining the approximate *convex hull* of a shape, and determining the skeleton by zone of influence. Thickening is normally only applied to binary images, and it produces another binary image as output.

The thickening operation is calculated by translating the origin of the strel to each possible pixel position in the image, and at each such position, comparing it with the underlying image pixels. If the foreground and background pixels in the structuring element exactly match foreground and background

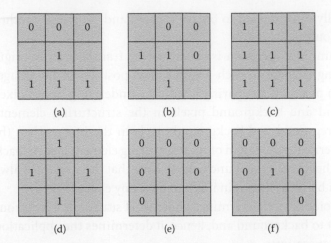

FIGURE 10.6 Some strels used in the thinning operation. (a) A piercing edge from bottom; (b) an angular piercing edge; (c) simply finds the boundary of a binary object; that is, it deletes any foreground points that do not have at least one neighboring background point (note that the detected boundary is 4-connected); (d) does the same thing but produces an 8-connected boundary; (e) and (f) are used for pruning, that is, reapplying thinning until convergence.

pixels in the image, then the image pixel under the origin of the structuring element is set to foreground (one). Otherwise, it is left unchanged. Note that the strel must always have a zero or a blank at its origin if it is to have any effect. The choice of strel determines in what situations a background pixel will be set to foreground and, hence, it determines the application for the thickening operation. Two typical strels are shown in Figure 10.8.

Figure 10.9 shows the application of this operator to a real image.

10.8 SKELETONIZATION/MEDIAL AXIS TRANSFORM

Skeletonization is a process for reducing foreground regions in a binary image to a skeletal remnant that largely preserves the extent and connectivity of the original region while throwing away most of the original foreground pixels. Essentially, in each iteration, some pixels are systematically removed without causing any hole or break to be created. To see how this works, imagine that the foreground regions in the input binary image are made of some uniform slow-burning material. Light fires simultaneously at all points along the boundary of this region, and watch the fire move into the interior. At points where the fire traveling from two different

FIGURE 10.7 Application of the thinning operator. (a) Resulting image after three iterations, (b) resulting image segment after one iteration, (c) actual values for (b).

boundaries meets itself, the fire will extinguish itself; the points at which this happens form the so-called *quench line*. This line is the skeleton. According to this definition, it is clear that the thinning produces a sort of skeleton.

The terms *medial axis transform* (MAT) and *skeletonization* are often used interchangeably but with the slight distinction that the skeleton is simply a binary image showing the simple skeleton, whereas MAT is a gray-level image where each point on the skeleton has an intensity that represents its distance to a boundary in the original object. The skeleton/

(a)   (b)

FIGURE 10.8 Structuring elements for determining the convex hull using thickening. During each iteration of the thickening, each element should be used in turn, and then in each of their 90° rotations, giving eight effective structuring elements in total. The thickening is continued until no further changes occur, at which point the convex hull is complete.

MAT can be produced in two main ways. The first is to use some kind of morphological thinning that successively erodes away pixels from the boundary (while preserving the end points of line segments) until no more thinning is possible, at which point what is left approximates the skeleton.

As with thinning, slight irregularities in a boundary will lead to spurious spurs in the final image that may interfere with recognition processes based on the topological properties of the skeleton. Despurring or pruning can be carried out to remove spurs of less than a certain length, but this is not always effective because small perturbations in the boundary of an image can lead to large spurs in the skeleton. Figure 10.10 shows an example of the application of the skeletonization operation.

10.9 ALGORITHMIC ACCOUNT

The basic morphological operations are fundamentally a masking process with a specific decision associated with it. For instance, in the case of erosion, once the strel mask is satisfied, that is, conditions such as $\forall (image(i,j) \wedge strel(i,j)) = 1$ —that is, whether there is a match on the image with the strel of erosion (i.e., all 1's in typically 3×3 settings)—then certain pixels, including the center pixel, are made zeros. This algorithmic approach is true for any other operator too. Figure 10.11(a) puts it in a graphical context. For the iterative operators, or repeated (more than once) applications of the operators, the procedure is shown in Figure 10.11(b).

10.10 MATLAB® CODE

In this section, the MATLAB code used to generate the images in Figures 10.1–10.5, and 10.7–10.9 is provided.

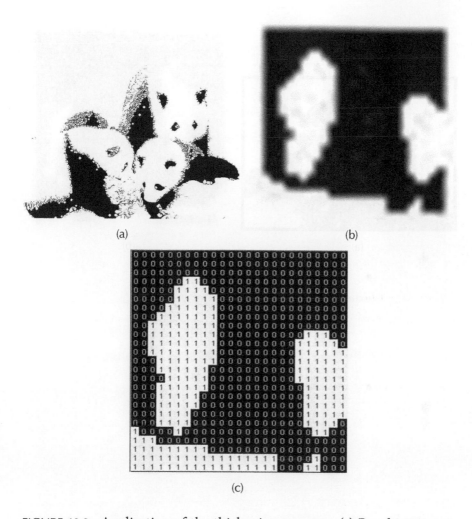

FIGURE 10.9 Application of the thickening operator. (a) Resulting image after three iterations, (b) resulting image segment after one iteration, (c) actual values for (b).

```
close all
clear all
x=imread('panda2.jpg');
figure ; imshow(x)
xb = im2bw(x,0.2) ;
figure ; imshow(xb)
x1=xb(566:590,882:906);
figure ; imshow(x1)

% Erosion
```

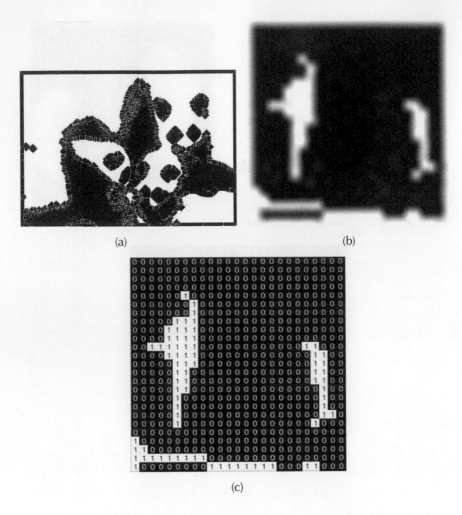

FIGURE 10.10 Application of the skeletonization operator. (a) Resulting image after three iterations, (b) resulting image segment after one iteration, (c) actual values for (b).

```
y1 = bwmorph(xb,'erode',5) ;
z1 = bwmorph(x1,'erode') ;
figure ; imshow(y1)
figure ; imshow(z1)

% Dilation
y2 = bwmorph(xb,'dilate',5) ;
z2 = bwmorph(x1,'dilate') ;
figure ; imshow(y2)
figure ; imshow(z2)
```

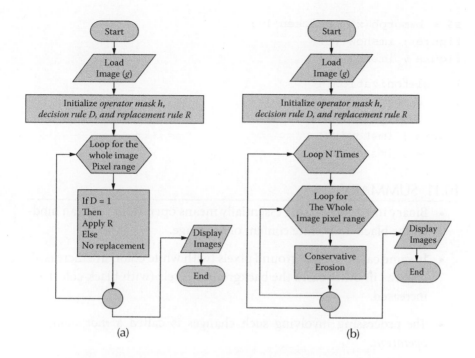

FIGURE 10.11 Procedure for iterative operators. (a) Algorithm for morphological operators, (b) iterative application of an operator to achieve, say, skeletonization.

```
% Opening
y3 = bwmorph(xb,'open',5) ;
z3 = bwmorph(x1,'open') ;
figure ; imshow(y3)
figure ; imshow(z3)

% Closing
y4 = bwmorph(xb,'close',5) ;
z4 = bwmorph(x1,'close') ;
figure ; imshow(y4)
figure ; imshow(z4)

% Thinning
y5 = bwmorph(xb,'thin',5) ;
z5 = bwmorph(x1,'thin') ;
figure ; imshow(y5)
figure ; imshow(z5)

% Thickening
y6 = bwmorph(xb,'thicken',5) ;
```

```
z6 = bwmorph(x1,'thicken') ;
figure ; imshow(y6)
figure ; imshow(z6)

% Skeletonization
y7 = bwmorph(xb,'skel',20) ;
z7 = bwmorph(x1,'skel') ;
figure ; imshow(y7)
figure ; imshow(z7)
```

10.11 SUMMARY

- Binary image processing essentially means operations that can modify the black-and-white content of an image.

- In some cases, the foreground pixels (with white color) are increased, whereas in other cases the background pixels (with black color) are increased.

- The processing involving such changes is called a *morphological operation*.

- The morphological operators dilation, closing, and thickening will increase the foreground content but in slightly varying degrees.

- Similarly, the morphological operators erosion, opening, and thinning will increase the background content but in varying degrees.

- If erosion is applied iteratively, keep in mind to avoid an opening or hole formation in the foreground objects, then it will converge to a final and most simplified connected version of the source image that is called the *skeleton* of the image.

- Morphological operations are extremely useful in isolating certain regions of interest of the image for further processing, or in removing or increasing certain types of pixels in certain neighborhoods, or in getting the skeleton of the objects present in the image.

10.12 EXERCISES

1. In the usual erosion operation, the strel is moved pixel by pixel as in convolutional shifting. Modify this procedure for the nonoverlapping mode of erosion, in which the mask is shifted to a new location on the image with overlapping with any other previously used

areas. Apply this process on the panda image, and compare with Figure 10.3.

2. Repeat Q #1 for dilation.

3. Use the Sobel operators of G_x and G_y (refer to Chapter 9 for details) as two strels for the thinning operation. First use G_x and then G_y. Comment on the resulting image.

4. Can a combination of the morphological operators discussed in this chapter be used to find edges in the image? If so, what is the sequence? If not, explain why not.

5. Use a noisy grayscale image with salt-and-pepper noise of default strength in MATLAB. Convert this image to binary, first using default settings in MATLAB. Then apply erosion, dilation, thinning, and thickening operators to the image three times each. Comment on the noise removal or failure to remove with respect to the signal-to-noise ratio (SNR) of the improved image.

CHAPTER 11

Image Encryption and Watermarking

11.1 INTRODUCTION

DURING THE LAST TWO decades, the issue of enhancing privacy in transmitted data has attracted the attention of many researchers from different fields, and many complementary methods have been suggested in order to improve privacy. As more and more information is being transferred electronically, the issue of interest here is that if someone accesses all of the data, he or she has the potential to copy it and retransmit it to unauthorized parties. This is because of the nature of digital data, which is easily reproducible. Usually, this topic is covered together with image encryption, cryptography, or watermarking. The focus in this chapter is on the watermarking technique. This technique does not distort the image but embeds some hidden information, which is like a secret signature that can be accessed by authorized users before copying it or retransmitting it. Watermarking was first used for copyright protection, but applications of watermarking have recently been expanded to content verification, authentication, covert communications, and information retrieval.

In some cases, the watermarking is done in such a way that the actual image is distorted in one way or the other. This could vary from including a specific text or logo appearing on the image or distortion to the extent that the actual image just does not appear the same way. Which of these is to be used in a particular situation depends on the level of security being imposed on the image data. If the data is confidential and cannot be allowed to be viewed by unauthorized users, it should be encrypted in such a way that the shape of the image is distorted beyond simple reconstruction logic. Thus, only authorized users can reverse the process, by applying the decoding procedures (inverse of the encoding technique used in that particular scenario). There are many techniques used in this context, and several of them cannot be fully revealed because of their applications in the defense sector.

On the other hand, when the intent is to just have a signature in the image so that if and when this image is reproduced by unauthorized users, then image encryption can be utilized. Failure to decode the embedded signature at any point would prove that the user was unauthorized, because an authorized user will be able to remove the embedded signature. There are different types of encryption techniques available in the literature, but watermarking alone is discussed in this chapter. Again, only a conceptual guideline is provided in this chapter due to the sensitive nature of the technique.

11.2 WATERMARKING METHODOLOGY

Digital watermarking is a method that has received a lot of attention in the past few years. A digital watermark can be described as a visible, or preferably invisible, identification code that is permanently embedded in the data. It remains present within the data after any decryption process. A general definition of watermarking is: "Hiding a secret message or information within an ordinary message and its extraction at the destination."

Watermarking is complementary to encryption, as it allows some protection of the data after decryption. As we know, the encryption procedure aims at protecting the image (or other kind of data) during its transmission. Once decrypted, the image is not protected anymore. By adding a watermark, we confer a certain degree of protection on the image (or on the information that it contains) even after the decryption process has taken place. The goal is to embed some information in the image without affecting its visual content. In the copyright protection context, watermarking is used to add a key in the multimedia data that authenticates the legal copyright holder, and that cannot be manipulated or removed without changing the data in a way that negates any commercial value.

The rapid expansion of the Internet in the past years has rapidly increased the public availability of digital data such as audio, images, and video. The problem of protecting multimedia information is becoming more and more important, and many copyright owners are concerned about illegal duplication of their data or work. The manner in which a user accesses the information differs, perhaps through secure hardware, and this allows one form of copyright protection. An example of this form of security is a DVD player. Instead, watermarking works by hiding data within an image, so that an unauthorized user will retransmit this hidden information, or watermark. Such a scenario is shown in Figure 11.1.

As can be seen, if Mr. B tries to resend this information without authorization, then Mr. A can prove the origin of the data. This type of watermark, known as a *robust watermark*, remains with the data as long as the data is still recognizable. If Mr. B degrades the data to such an extent that it becomes meaningless, then it is not reasonable to expect the watermark to still be obtainable. On the other hand, if Mr. B is an authorized user but not an authorized distributor, he should also receive the key to dewatermark the image before using or he should receive the dewatermarked image.

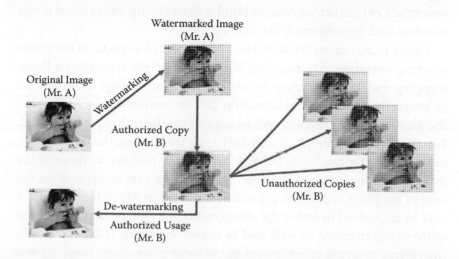

FIGURE 11.1 (See color insert following Page 204.) Concept of watermarking.

11.3 BASIC PRINCIPLE OF WATERMARKING

Because digital watermarking is a technique that allows for the secret embedding of information in host data, an image watermarking scheme should at least meet the following requirements:

- *Transparency:* The embedded watermark should be perceptually invisible
- *Robustness:* The embedded watermark should not be erased by any attack that maintains an acceptable host image quality.

The trade-off between transparency and robustness is one of the most important issues in image watermarking. In terms of the embedding domain, watermarking algorithms are mainly divided into two groups: spatial domain methods, which embed the data by directly modifying the pixel values of the original image, and transform domain methods, which embed the data by modulating the transform domain coefficients. The most commonly used transforms for digital watermarking are the DFT (discrete Fourier transform), DCT (discrete cosine transform), and DWT (discrete wavelet transform). In general, spatial domain methods have good computing performance, and transform domain methods have high robustness. In terms of the extracting scheme, watermarking algorithms are also divided into two groups: blind and nonblind watermarking. In nonblind watermarking, the original image is necessary for the watermark extraction, whereas in blind watermarking, the original image is not needed for watermark extraction.

Figure 11.2(a) shows the actual procedure needed to perform the watermarking operation. The mapping block $F' = F + gW$ represents a linear mapping method of merging the watermark binary image into the RGB or grayscale images. The relationship can be modified as needed. Also, the location where mapping will be applied can be technique dependent. For instance, with the DCT- and DFT-type transforms, the most significant coefficients can be sorted out in the same number as those of the pixels in the watermark, and then the mapping can be applied. In the case of the DWT, a different approach is utilized in which the watermark may be embedded in one of the components, etc. The recovery process is quite straightforward as well, and is shown in Figure 11.2(b). Basically, the reverse mapping is performed in the same place where mapping was performed in the first phase.

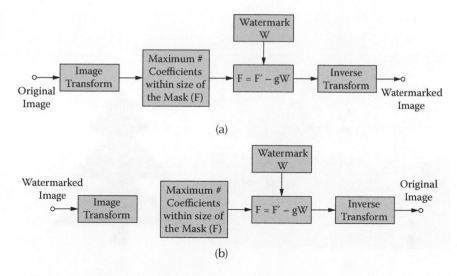

FIGURE 11.2 Watermarking and recovery. (a) Watermarking process, (b) its inverse process.

Figure 11.3 shows the results of application of a simple binary watermark to an RGB image and its recovery. Note the modifications made to the RGB image by the watermark, which are prominent on the top side of the watermarked image, as well as the absolute difference calculated in Figure 11.3(d).

11.4 PROBLEMS ASSOCIATED WITH WATERMARKING

The idea of watermarking seems to be very nice and robust at first glance; however, it runs into a set of constraints and implementation issues that limit the robustness of the technique. The difficulty of inserting a robust watermark is that it may be possible for an unauthorized user to remove the watermark through ignorance or, more likely, by intention. Suppose we take an image, Figure 11.4, and embed a watermark in it. A watermark must modify the information contained in the original image, so we know that pixels within the image must be changed. This change is the watermark itself and identifies the owner. If the watermark is visible, then it can be removed by filtering so that there are no noticeable distortions. This imposes the requirement that a robust watermark should be invisible. Figure 11.4 shows the effect of adding Gaussian noise to the system, which causes pixel information to be modified, but recovery is still possible because the information was not distorted to a great extent.

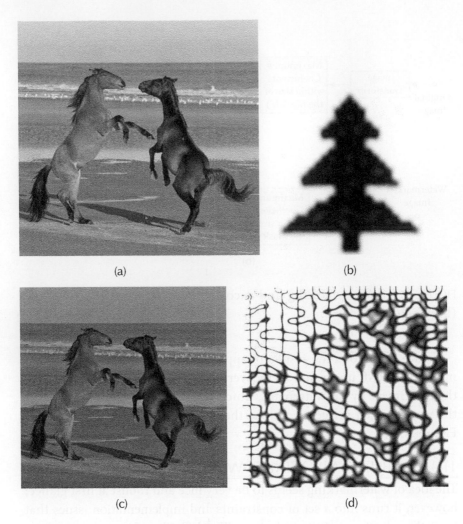

FIGURE 11.3 Application of a binary watermark on an RGB image. (a) Original image, (b) watermark, (c) watermarked image, (d) absolute difference between the images in (a) and (c).

If an unauthorized user performs (lossy) compression on the image, such as JPEG compression with the aim of reducing the file size, the quality of the image is also reduced. The consequence of this is that information stored in the pixels may also be eroded. This will also change the embedded watermark to the extent that it may not be recognizable.

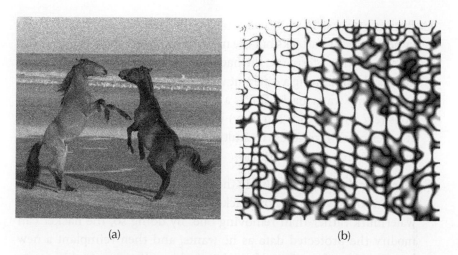

(a)　　　　　　　　　　　　　　　　　(b)

FIGURE 11.4　Application of a binary watermark on a noisy RGB image. (a) Noisy watermarked image, (b) absolute difference of the images.

11.4.1 Attacks on Watermarks

There are several kinds of malicious attacks that result in partial or even total destruction of the embedded identification key and for which more advanced watermarking schemes should be employed. Trickier attacks involve rotations of the image, warping, and scaling of the image. Also possible is a cropping attack, which removes rows and columns from the image to form a new shape. The difficulty with this attack is that it affects the synchronization of the image. In order to detect a watermark, there needs to be some form of comparison of the signal against a benchmark. If the signal is close enough to the expected form, then the watermark is detected. However, if synchronization is lost, then the watermark will not be detected. These attacks can be classified as follows:

Active attacks: Here, the hacker tries deliberately to remove the watermark or simply make it undetectable. This is a big issue in copyright protection, fingerprinting, or copy control, for example.

Passive attacks: In this attack, the attacker does not try to remove the watermark but simply attempts to determine if a given mark is present. As the reader will understand, protection against passive attacks is of the utmost importance in covert communications, where even knowledge of the presence of a watermark is often more than one wants to grant.

Collusion attacks: In collusive attacks, the goal of the hacker is the same as in active attacks but the method is slightly different. In order to remove the watermark, the hacker uses several copies of the same data, each containing a different watermark, to construct a new copy without any watermark. This is a problem in fingerprinting applications (e.g., in the film industry) but is not widespread, because the attacker must have access to multiple copies of the same data and the number needed can be critical.

Forgery attacks: This is probably the main concern in data authentication. In forgery attacks, the hacker aims at embedding a new, valid watermark rather than removing one. By doing so, the hacker can modify the protected data as he wants, and then reimplant a new key to replace the original key, thus making the corrupted image to appear genuine.

11.4.2 What Can Be Done?

Each attack is dealt with by a different approach, and the development of a robust technique for image processing remains an open problem. However, some of the commonly used methods are now listed briefly.

Redundant watermarks: Here, the watermark is encoded into more than one known place in the target image. This can be used against cropping and noise attacks.

Multiple watermarks: This implies embedding more than one watermark in the image at known positions. If the positions and sizes of the watermarks are selected in such a way that they can cover both even and odd number spaces, then a compression attack can be countered.

Symmetric watermarks: The binary watermark must be selected as a symmetric image so that the rotation of the image will affect this watermark significantly. The positioning of such a watermark will also be important to counter a rotation attack.

Transformed watermarks: Instead of a simple image being embedded in the transformed image, a transformed version of this watermark will distribute the information within the whole image more uniformly and, thus, the image can be made more robust to certain attacks.

Nonlinear mapping: Instead of using simple linear or affine mapping, advanced mappings can be utilized, making it more difficult for an attacker to isolate the watermark.

11.5 ALGORITHMIC ACCOUNT

A typical methodology for coding the watermarks is shown in Figure 11.5.

11.6 MATLAB® CODE

In this section, the MATLAB code used to generate the images in Figures 11.3 and 11.4 is presented.

```
clear all
close all
x = double(imread('horses.jpg'));
figure ; imshow(x*0.003) ;
y = x ;
```

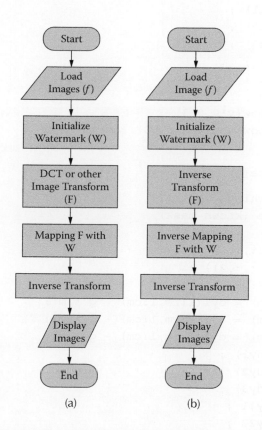

(a)   (b)

FIGURE 11.5 Methodology for coding watermarks. (a) Watermarking procedure, (b) dewatermarking procedure.

```
% Water marking
x1 = x(:,:,1) ;
x2 = x(:,:,2) ;
x3 = x(:,:,3) ;
dx1 = dct2(x1) ; dx11 = dx1 ;
dx2 = dct2(x2) ; dx22 = dx2 ;
dx3 = dct2(x3) ; dx33 = dx3 ;
load m.dat % binary mask for watermarking
g = 10 ; % Coefficient of watermark's strength
[rm,cm] = size(m) ;
dx1(1:rm,1:cm) = dx1(1:rm,1:cm) + g * m ;
dx2(1:rm,1:cm) = dx2(1:rm,1:cm) + g * m ;
dx3(1:rm,1:cm) = dx3(1:rm,1:cm) + g * m ;
y1 = idct2(dx1) ;
y2 = idct2(dx2) ;
y3 = idct2(dx3) ;
y(:,:,1) = y1 ;
y(:,:,2) = y2 ;
y(:,:,3) = y3 ;
figure ; imshow(y*0.003)
figure ; imshow(abs(y-x)*100) %comparison

z = y ;
[r,c,s] = size(z) ;

% De-watermarking
% Clean image (known mask)
y = z ;
dy1 = dct2(y(:,:,1)) ;
dy2 = dct2(y(:,:,2)) ;
dy3 = dct2(y(:,:,3)) ;
dy1(1:rm,1:cm) = dy1(1:rm,1:cm) - g * m ;
dy2(1:rm,1:cm) = dy2(1:rm,1:cm) - g * m ;
dy3(1:rm,1:cm) = dy3(1:rm,1:cm) - g * m ;
y11 = idct2(dy1) ;
y22 = idct2(dy2) ;
y33 = idct2(dy3) ;
yy(:,:,1) = y11 ;
yy(:,:,2) = y22 ;
yy(:,:,3) = y33 ;
figure ; imshow(yy*0.003)
figure ; imshow(abs(yy-x)*10000) %comparison showing all
black image for no difference b/w yy and x
```

```
% Noisy image
n = 100 ;
y1 = z(:,:,1); y1 = y1 + rand(r,c)*n;
y2 = z(:,:,2); y2 = y2 + rand(r,c)*n;
y3 = z(:,:,3); y3 = y3 + rand(r,c)*n;
y(:,:,1) = y1 ;
y(:,:,2) = y2 ;
y(:,:,3) = y3 ;
figure ; imshow(y*0.003)
dy1 = dct2(y1) ;
dy2 = dct2(y2) ;
dy3 = dct2(y3) ;
dy1(1:rm,1:cm) = dy1(1:rm,1:cm) - g * m ;
dy2(1:rm,1:cm) = dy2(1:rm,1:cm) - g * m ;
dy3(1:rm,1:cm) = dy3(1:rm,1:cm) - g * m ;
y11 = idct2(dy1) ;
y22 = idct2(dy2) ;
y33 = idct2(dy3) ;
yy(:,:,1) = y11 ;
yy(:,:,2) = y22 ;
yy(:,:,3) = y33 ;
figure ; imshow(yy*0.003)
figure ; imshow(abs(yy-x)*0.0001) %comparison
```

11.7 SUMMARY

- Watermarking is embedding additional imagelike information in a target image.

- This is usually done in order to authenticate the image and to detect if an unauthorized copy has been made.

- The process of embedding this watermark is implemented as a linear mapping of the watermark image into the image under study. Although other mappings are also possible, only linear mapping is considered in this chapter.

- Usually, this mapping is done after transforming the image into a frequency-related space.

- Commonly used transforms for this purpose are the DCT and DWT.

- The watermark is mapped with the significant coefficients of the transformed image and then inverse-transformed to retrieve the image representation.

- Modifications in a watermarked image, such as addition of noise, rotation, cropping, and compression, can increase the watermark distortion so much that it may not be recoverable.

- Robustness can be achieved in the watermarked images through redundant watermarks, multiple watermarks, and mapping the watermarks using transformed binary images rather than the image itself.

- This is still an open area of research, and newer techniques are discovered every now and then.

11.8 EXERCISES

1. In Figure 11.4, a noise attack has been simulated. Repeat the same exercise with rotation attack, that is, after embedding the watermark, rotate the whole image by 450° and then repeat the same operation with −450°. Then try to recover the watermark.

2. Repeat Q #1 for 10% cropping from all sides.

3. Repeat Q #1 for JPEG compression. In this case, save the image as a 70% compressed JPEG image, then reopen it and apply the procedure to recover the watermark.

4. Apply the methodology that resulted in Figure 11.3, but use the DWT instead of the DCT. Refer to Section 5.3 for details of the transformation. Embed the watermark with HH coefficients.

5. Repeat Q #4 but embed the watermark with HL, LH, and HH components in the top-left, top-right, and bottom-right corners, respectively. Refer to Section 5.3 for explanation of these terms.

CHAPTER 12

Image Classification and Segmentation

12.1 INTRODUCTION

CLASSIFICATION, IN GENERAL, REFERS to clustering or grouping data items into similar sets. This information is often useful in the analysis step for any signal/data processing system. Image classification is similar to general data classification, but it may be different depending on the application in which it is used. For instance, in a satellite image, it might be required to isolate the fields from roads or plains from localities. This information may be needed for visual inspection for planning purpose only and, hence, it would be sufficient to redraw these artifacts in various colors representing each class with a different but meaningful color. On the other hand, the same information may be needed by an unmanned airplane to carry out a certain military mission, for which purpose a knowledge of the centers or certain coordinates on these classes would be needed rather than colored representations. Accordingly, the task of image classification becomes more complicated.

A human analyst attempting to classify features in an image uses the elements of visual interpretation to identify homogeneous groups of pixels that represent various features for the classes of interest. Digital image classification uses the spectral information represented by digital numbers in one or more spatial or spectral bands, and attempts to classify each individual pixel based on this information. The resulting classified image comprises a mosaic of pixels, each of which belongs to a particular class, and is essentially a thematic "map" of the original image.

When talking about classes, we need to distinguish between *information* classes and *spatial/spectral* classes. Information classes are those categories of interest that the analyst is actually trying to identify in the imagery, such as different kinds of crops, different forest types or tree species, different geologic units or rock types, etc. Spatial/spectral classes are groups of pixels that are uniform (or near-similar) with respect to their brightness values in the different spectral channels of the data. The objective is to match these classes in the data to the information classes of interest. Rarely is there a simple one-to-one match between these two types of classes. Rather, unique spatial/spectral classes may appear that do not necessarily correspond to any information class of particular use or interest to the analyst. It is the analyst's job to decide on the utility of the different classes and their correspondence to useful information.

Common classification procedures can be broken down into two broad subdivisions, based on the method used.

12.1.1 Supervised Classification

In this classification method, the analyst identifies in the imagery homogeneous representative samples of the different surface cover types (information classes) of interest. These samples are referred to as *training areas*. The selection of appropriate training areas is based on the analyst's familiarity with the geographical area and their knowledge of the actual surface cover types present in the image. Thus, the analyst is "supervising" the categorization of a set of specific classes. The numerical information in all spectral bands for the pixels comprising these areas are used to "train" the computer to recognize spectrally similar areas for each class. The computer uses a special program or algorithm (of which there are several variations) to determine the numerical "signatures" for each training class. Once the computer has determined the signatures for each class, each pixel in the image is compared to these signatures and labeled as the class it most closely "resembles" digitally.

12.1.2 Unsupervised Classification

This method in essence reverses the supervised classification process. Spatial/spectral classes are grouped first, based solely on the numerical information in the data, and are then matched by the analyst to information classes (if possible). Programs, called *clustering algorithms*, are used to determine the natural (statistical) groupings or structures in the data. Usually, the analyst specifies how many groups or clusters are to be looked for in the data. In addition to specifying the desired number of classes, the analyst may also specify parameters related to the separation distance among the clusters and the variation within each cluster. The final result of this iterative clustering process may result in some clusters that the analyst will want to subsequently combine, or clusters that should be broken down further, each of these requiring a further application of the clustering algorithm. Thus, unsupervised classification is not completely without human intervention. However, unlike a supervised classification, it does not start with a predetermined set of classes.

12.2 GENERAL IDEA OF CLASSIFICATION

Classification includes a broad range of decision-theoretic approaches to the identification of images (or parts thereof). All classification algorithms are based on the assumption that the image in question depicts one or more features (e.g., geometric parts in the case of a manufacturing classification system, or spectral regions in the case of remote sensing) and that each of these features belongs to one of several distinct and exclusive classes. The classes may be specified a priori by an analyst (as in *supervised classification*) or automatically clustered (i.e., as in *unsupervised classification*) into sets of prototype classes, where the analyst merely specifies the number of desired categories. (Classification and *segmentation* have closely related objectives, as the former is another form of component labeling that can result in segmentation of various features in a scene.)

Image classification analyzes the numerical properties of various image features and organizes data into categories. Classification algorithms typically employ two phases of processing: *training* and *testing*. In the initial training phase, characteristic properties of typical image features are isolated and, based on these, a unique description of each classification category, that is, training class, is created. In the subsequent testing phase, these feature-space partitions are used to classify image features.

The description of training classes is an extremely important component of the classification process. In supervised classification, *statistical* processes (i.e., based on an a priori knowledge of probability distribution functions) or *distribution-free* processes can be used to extract class descriptors. Unsupervised classification relies on *clustering* algorithms to automatically segment the training data into prototype classes. In either case, the motivating criteria for constructing training classes are that they are

- *Independent*: A change in the description of one training class should not change the value of another.

- *Discriminatory*: Different image features should have significantly different descriptions.

- *Reliable*: All image features within a training group should share the common definitive descriptions of that group.

12.3 COMMON INTENSITY-CONNECTED PIXEL: NAÏVE CLASSIFIER

Connected components labeling scans an image and groups its pixels into components based on pixel connectivity; that is, all pixels in a connected component share similar pixel intensity values and are in some way connected with each other. Once all groups have been determined, each pixel is labeled with a gray level or a color (color labeling) according to the component it was assigned to. Extracting and labeling of various disjoint and connected components in an image is central to many automated image analysis applications.

- If all four neighbors are 0, assign a new label to p, else

- if only one neighbor has $V = \{1\}$, assign its label to p, else

- if more than one neighbor has $V = \{1\}$, assign one of the labels to p and make a note of the equivalences.

After completing the scan, the equivalent label pairs are sorted into equivalence classes and a unique label is assigned to each class. As a final step, a second scan is made of the image, during which each label is replaced by the label assigned to its equivalence classes. For display, the labels might be different gray levels or colors.

As the first example of such a procedure, consider Figure 12.1. The figure shows the RGB image of the horses with sky, land, water, and the horses as the four main classes visible in the image. A very naïve method of classification would be to use the grayscale ranges that represent these classes. Obviously, this is not a very good classifier because different gray shades will be found for different objects not related to each other.

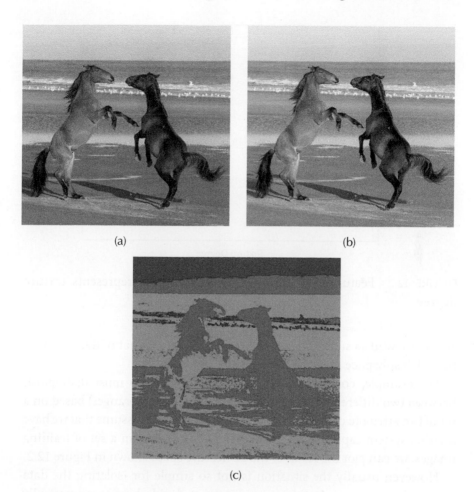

(a)

(b)

(c)

FIGURE 12.1 (See color insert following Page 204.) Image classification using the grayscale ranges only. (a) Original RGB image, (b) grayscale version of (a), (c) segmented image.

12.4 NEAREST NEIGHBOR CLASSIFIER

A convenient way of building a parametric description of the kind mentioned in the previous section is by using a feature vector $[\theta_1 \quad \theta_2 \quad \cdots \quad \theta_n]$, where n is the number of attributes that describe each image feature and training class. This representation allows us to consider each image feature as occupying a point, and each training class as occupying a subspace (i.e., a representative point surrounded by some spread, or deviation) within the n-dimensional classification

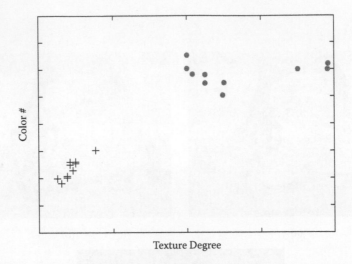

FIGURE 12.2 Feature space: + represents color #, o represents texture degree.

space. Viewed as such, the classification problem is that of determining to which subspace class each feature vector belongs.

For example, consider an application in which we must distinguish between two different types of objects (e.g., apples and oranges) based on a set of two attribute classes (e.g., *color* and *texture*). If we assume that we have a vision system capable of extracting these features from a set of training images, we can plot the result in the 2D feature space shown in Figure 12.2.

However, usually the situation is not so simple for isolating the data into various groups. At this point, we must decide how to numerically partition the feature space so that if we are given the feature vector of a test object, we can determine, quantitatively, to which of the two classes it belongs. One of the simplest (although not the most computationally efficient) techniques is to employ a supervised, distribution-free approach known as the *minimum (mean) distance classifier or nearest neighbor classifier*. This technique is now described.

Minimum (mean) distance classifier. Among the various methods of supervised statistical pattern recognition, the nearest neighbor rule achieves consistently high performance, without a priori assumptions about the distributions from which the training examples are drawn. It involves a training set of both positive and negative cases. A new sample is classified by calculating the distance to the nearest training case; the sign of that point then determines the classification of the sample. The *k*-NN classifier extends this idea by taking the k nearest points and assigning the sign of the major-ity. It is common to assign small and odd numbers to k to break ties (typi-cally 1, 3, or 5). Larger k values help reduce the effects of noisy points within the training data set, and the choice of k is often obtained through cross-validation.

12.4.1 Mechanism of Operation

Suppose that each training class is represented by a prototype (or *mean*) vector. This is particularly useful when classifying based on colors because for a selected area, the average color would be a good representation of the neighborhood. Such a mean is given as

$$\mu_n = \frac{1}{N_n}\sum_{x\in\omega_n} x \qquad for \qquad n=1,2,...,M, \tag{12.1}$$

where N_n is the number of training pattern vectors from class ω_n. In a color-based classification system, usually an area is first selected for a par-ticular class and all of its enclosed pixels, x, are then averaged to get the representative mean. Based on this, we can assign any such given pattern to the class of its closest prototype by determining its proximity to each μ_n. If Euclidean distance is our measure of proximity, then the distance to the prototype is given by

$$D_n(x)=\|x-m_n\| \qquad for \qquad j=1,2,...,M, \tag{12.2}$$

which is equivalent to computing

$$D_n(x)=x^T m_n - \frac{m_n^T m_n}{2} \qquad for \qquad n=1,2,...,M, \tag{12.3}$$

and assigning x to class ω_n if $D_n(x)$ yields the smallest value. The *decision boundary* that separates class ω_i from ω_j is given by values for x for which

$$D_i(x)-D_j(x)=0. \tag{12.4}$$

Figure 12.3 shows the application of this algorithm to the sample image.

In practice, the minimum (mean) distance classifier works well when the distance between means is large compared to the spread (or random-ness) of each class with respect to its mean. It is simple to implement and is expected to give an error rate within a factor of two of the ideal error rate, obtainable with the statistical, supervised *Bayes' classifier*. The Bayes'

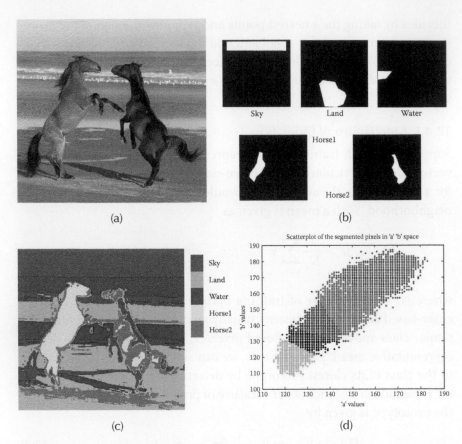

FIGURE 12.3 (See color insert following Page 204.) Application of nearest neighbor algorithm to classify the given image into its classes. (a) Sample image, (b) selected areas as training classes, (c) segmented image, (d) class boundaries.

classifier is a more "informed" algorithm as the frequencies of occurrence of the features of interest are used to aid the classification process. Without this information, the minimum (mean) distance classifier can yield biased classifications. This can be best combated by applying training patterns at the natural rates at which they arise in the raw training set.

12.5 UNSUPERVISED CLASSIFICATION

The second method employed for this classification is called *unsupervised segmentation*. In general, unsupervised clustering techniques are used less frequently, as the computation time required for the algorithm to learn

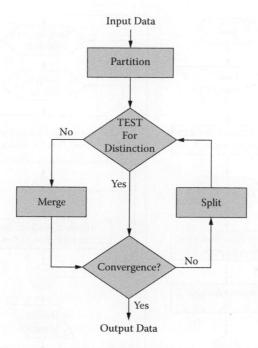

FIGURE 12.4   General clustering algorithm.

a set of training classes is usually prohibitive. However, in applications where the features (and relationships between features) are not well understood, clustering algorithms can provide a viable means of partitioning a sample space.

A general clustering algorithm is based on a *split and merge* technique, as shown in Figure 12.4. Using a *similarity measure* (e.g., the dot product of two vectors, the weighted Euclidean distance, etc.), the input vectors can be partitioned into subsets, each of which should be sufficiently distinct. Subsets that do not meet this criterion are merged. This procedure is repeated on all of the subsets until no further splitting of subsets occurs or until some *stopping criteria* are met.

12.6 ALGORITHMIC ACCOUNT

Classification is such a wide-ranging field that a description of all the algorithms could fill several volumes of text. We have already discussed a common supervised algorithm in the previous section, and Figure 12.5 shows the nearest neighbor algorithm method.

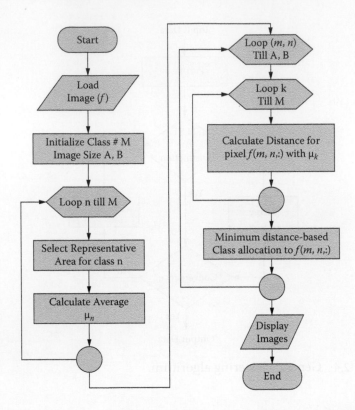

FIGURE 12.5 Nearest neighbor classifier for image classification and segmentation.

12.7 MATLAB® CODE

In this section, the actual MATLAB code used to generate the images in Figures 12.1 and 12.3 is presented. Starting with the grayscale classification method of Figure 12.1, the required code is as follows:

```
close all
clear all
x = imread('horses.jpg') ;
figure; imshow(x)
xg = rgb2gray(x) ;
figure ; imshow(xg)
cl1 = [191 220] ; % sky
c1 = [255 0 0 ] ;
cl2 = [ 131 190 ] ; % land
c2 = [ 0 255 0 ] ;
```

```
cl3 = [0 130 ] ; % horse
c3 = [ 255 0 255 ] ;
cl4 = [ 221 255 ] ; % water
c4 = [ 0 0 255 ] ;
[r,c] = size(xg) ;
y = x ;

for i = 1 : r
    for j = 1 : c
        if (xg(i,j) >= cl1(1) && xg(i,j) <= cl1(2))
            cc = c1 ;
        else
            if (xg(i,j) >= cl2(1) && xg(i,j) <= cl2(2))
                cc = c2 ;
            else
                if (xg(i,j) >= cl3(1) && xg(i,j) <= cl3(2))
                    cc = c3 ;
                else
                    cc = c4 ;
                end
            end
        end

        for k = 1 : 3
            y(i,j,k) = cc(k) ;
        end
    end
end

figure
imshow(y)
```

Following is the code for Figure 12.3:

```
close all
clear all
```

```
% Read the source RGB image
x = imread('horses.jpg');
figure ; imshow(x)
[r,c,s] = size(x) ;
```

```
% Initialize storage for each sample region.
classes = { 'sky','land','water','horse1','horse2' };
nClasses = length(classes);
sample_regions = false([r c nClasses]);
```

```
% Select each sample region.
f = figure;
for count = 1:nClasses
 set(f, 'name', ['Select sample region for '
classes{count}] );
 sample_regions(:,:,count) = roipoly(x);
end
close(f);

% Display sample regions.

for count = 1:nClasses
 figure
 imshow(sample_regions(:,:,count))
 title(['sample region for ' classes{count}]);
end

% Convert the RGB image into an L*a*b* image

cform = makecform('srgb2lab');
lab_x = applycform(x,cform);

% Calculate the mean 'a*' and 'b*' value for each ROI area

a = lab_x(:,:,2);
b = lab_x(:,:,3);
color_markers = repmat(0, [nClasses, 2]);

for count = 1:nClasses
 color_markers(count,1) =
mean2(a(sample_regions(:,:,count)));
 color_markers(count,2) =
mean2(b(sample_regions(:,:,count)));
end

%% Step 3: Classify Each Pixel Using the Nearest
Neighbor Rule
% Each color marker now has an 'a*' and a 'b*' value.
You can classify each pixel
% in the |lab_x| image by calculating the Euclidean
distance between that
% pixel and each color marker. The smallest distance
will tell you that the
% pixel most closely matches that color marker. For
example, if the distance
```

```
% between a pixel and the red color marker is the
smallest, then the pixel would
% be labeled as a red pixel.

color_labels = 0:nClasses-1;
a = double(a);
b = double(b);
distance = repmat(0,[size(a), nClasses]);

% Perform classification
for count = 1:nClasses
 distance(:,:,count) = ( (a - color_markers(count,1)).^2 +
...
                        (b - color_markers(count,2)).^2 ).^0.5;
end

[value, label] = min(distance,[],3);
label = color_labels(label);
%clear value distance;

colors = [ 255 0 0 ; 0 255 0 ; 0 0 255 ; 255 255 0 ; 255
0 255 ] ;
y = zeros(size(x)) ;
l = double(label)+1 ;
for m = 1 : r
    for n = 1 : c
        y(m,n,:) = colors(l(m,n),:) ;
    end
end

figure ; imshow(y)

colorbar

% scatter plot for the nearest neighbor classification
purple = [119/255 73/255 152/255];
plot_labels = { 'k', 'r', 'g', purple, 'm', 'y'};

figure
for count = 1:nClasses
 plot(a(label==count-1),b(label==count-
1),'.','MarkerEdgeColor', ...
 plot_labels{count}, 'MarkerFaceColor',
plot_labels{count});
 hold on;
end
```

```
title('Scatterplot of the segmented pixels in ''a*b''
space');
xlabel('''a*'' values');
ylabel('''b*'' values');
```

12.8 SUMMARY

- Image classification and segmentation refers to finding and labeling areas in an image that bear some resemblance to each other.

- This can be used to isolate certain areas of interest from an image or segment them with similar colors to distinguish such areas within the image.

- Applications for such segmentation are found, in particular, in remote sensing and medical imaging and, in general, for any image processing system. For instance, the classification technique can be used to categorize roads from fields in a satellite image.

- The basis of classification is to find areas of similar characteristics and mark them.

- These characteristics are called the features or attributes of each class, and can be calculated in a variety of ways. In this chapter, only color grayscale-based features are presented.

- If the features are calculated based on some guideline applied to the source image in order to identify a class through this guideline, then such a classification technique is called supervised classification.

- If, on the other hand, the classification is done automatically based on some form of clustering, then it is called unsupervised classification.

- Unsupervised classification is usually more time consuming than the supervised method.

- A variety of techniques have been reported in the literature to perform these classifications, and they vary in their accuracy and speed of application.

12.9 EXERCISES

1. Repeat the example of Figure 12.1 by using the five classes of Figure 12.3.

2. Instead of using the average as the main feature of a class, use the median as an attribute, and repeat the example of Figure 12.3.

3. Repeat Q #2, using entropy as the main feature.

4. Repeat Q #2 for a compound feature set of mean, median, and entropy.

5. Repeat Q #1 for a compound feature set of mean, median, and entropy.

CHAPTER 13

Image-Based Object Tracking

13.1 INTRODUCTION

A<small>UTOMATED SEGMENTATION AND TRACKING</small> are two fundamental and, at the same time, highly challenging tasks in image analysis. Application areas are numerous: medical and biomedical image processing, industrial image analysis, earth sciences, geographical information services, forestry services, to name a few. Today's satellite-based military applications also make use of these techniques for surveillance and knowing the whereabouts of a particular target. Several image segmentation and tracking methods have been proposed, and these have evolved greatly over the past few decades for several applications. Given an application, for a beginner, choosing appropriate segmentation or tracking methods from among several dozens of the available techniques could be a truly overwhelming task.

13.2 METHODOLOGIES

To be able to track moving objects in a video sequence, two processing steps are needed:

1. Localization of the target

2. Estimation of the change in position of the target

Essentially, the problem is then to trace the object's movements throughout the sequence. This has been done in a number of ways by various researchers employing different methods. This chapter focuses on some of the basic techniques that can be utilized, and upon which more complicated algorithms and techniques can be built.

Looking at the tracking problem from an image processing perspective, there are certain issues that any technique must resolve in order to perform the desired functionality. These primarily include the following:

1. Background variations

2. Noise

3. Changing shape or color of the target

4. Multiple objects of the same shape or color

In this chapter, four methods are outlined for a simple experimental sequence of images. The sequence comprises one to three different-colored balloons being flown (with hand blows only) within viewfinder range of a fixed digital camcorder. The balloons appear in the video sequence with variable sizes owing to perspective motion. A major problem is shadows appearing in the scene because of a light source above the viewing area. The techniques presented here were capable of detecting the targeted balloon quite successfully for most of the frames in the sequence. Some of the sample frames are shown in the respective figures, as well as being presented in the MATLAB® code. The methods employed suffered severely in their precision due to shadows as well as background shading variations (which introduced the bloblike noise in the images). The main conclusion drawn from the findings is that a particular simple technique may work within a certain limited scenario only. For an advanced application, a combination of various techniques, or even superior techniques, must be used.

The four techniques presented here are

1. Background subtraction

2. Temporal frames differences

3. Cross-correlation method

4. Color-based segmented tracking

13.3 BACKGROUND SUBTRACTION

Background subtraction basically means removing the moving objects from the background of an image. The method makes use of two or more frames, one of which is a reference frame consisting of only the background environment. The reference frame is then subtracted from the other frames, so that if the environment is static—that is, nothing has been changed in the image—the difference will be zero. However, if something has changed in the video sequence, the difference between the reference frame and the other frames will not be zero. Given a nonzero difference, a change, which can be an object or any other moving artifact, has been detected. The sample reference image and moving objects in the foreground are shown in Figure 13.1. The image in Figure 13.1(a) shows

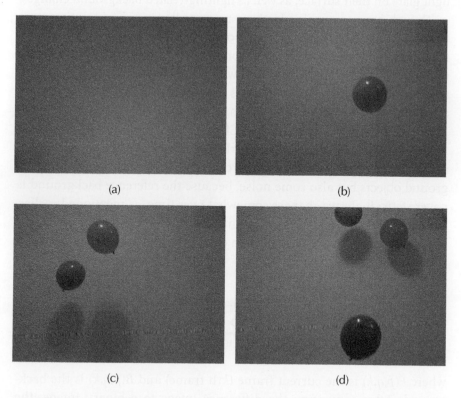

(a)

(b)

(c)

(d)

FIGURE 13.1 Sample frames obtained from the video sequence. (a) Background image, (b) one target only, (c) two objects with one being the target, (d) three objects with one being the target.

the blank background without any foreground artifacts in it. This is very useful information, and assuming that most of the background does not change, subtracting this from any subsequent image will reduce the information content the tracking algorithm needs to process.

The other images in Figure 13.1 represent the foreground artifacts (balloon), from which the red balloon is selected as the main target. Although the selection is completely random and any other balloon could have been selected as well, the shape of the artifact is what makes these images a bit easier to work with. The only change expected in them is their size, due to the distance from the camera. The symmetrical nature of the objects is a highly desirable characteristic that enables powerful techniques such as convolution- or correlation-based classification to be easily implemented.

The main problems were the inconsistent shadows of the balloons and the light glare on their surface, as well as lighting-related background changes.

Though it may seem simple, there are actually several subtraction steps to go through in order to achieve good results. The technique works well for a static environment with single or very few moving objects (including the target). Changes in background will cause the bloblike noise. In all of the following algorithms, this noise is removed at the end of the algorithm as a postprocessing procedure. This has been done here by using the morphological operations (discussed in Chapter 10).

The first step is, as described earlier, to take the current frame and subtract it from the reference background. The difference yields the foreground objects but also some noise, because the reference background is never static; it always changes somewhat over time, as mentioned earlier. One can also apply an M × M averaging filter in order to reduce this noise. The absolute difference between the current frame and the background frame is defined by

$$G_1(m,n,k) = \frac{1}{M^2} \left| \sum_{p=1}^{M-1} \sum_{q=1}^{M-1} F(p,q,k) - B(p,q,k) \right|, \tag{13.1}$$

where $F(p,q,k)$ is the current frame (kth frame) and $B(p,q,k)$ is the background. After converting the difference image to a binary image, the changes that have occurred will be highlighted in white against a solid black background. An example of this can be seen in Figure 13.2.

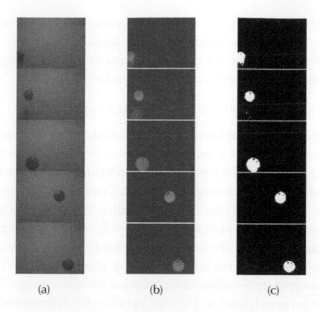

(a)   (b)   (c)

FIGURE 13.2   Object detection with background difference. (a) Original image sequence, (b) difference image in grayscale, (c) difference image after being converted to a binary image.

13.3.1 Artifacts

Artifacts often appear in the basic difference images described earlier as dots or groups of dots (bloblike noise). This is due to unwanted changes and movements in the images and acquisition, such as:

- Sudden changes in camera positions due to a gust of air or other vibratory motion

- Small background changes due to shadows or lighting fluctuations

- Changes in the foreground *not* due to the target's movement, including multiple moving objects

Artifacts might be interpreted as moving objects that do not belong to the background environment, even though they should be considered part of a reference environment. These problems occur when using only one static reference frame because the reference frame is never updated to include such changes to the background. The best way to avoid this would be to update the reference image over time using adaptive methods, but that is beyond the scope of this book. Thus, to reduce the impact of these artifacts, we need to subtract more than once between the frames. These subtractions must also be made in different ways so that we can accommodate the different kinds of artifacts presented earlier.

13.4 TEMPORAL DIFFERENCE BETWEEN FRAMES

The temporal difference between frames calculates the difference between subsequent frames. This is because if an object belonging to the background has been moved, it will be visible in every background difference image at the same position until it moves again. However, if we take the difference between, let us say, the current frame and the preceding one, where the moved chair first appears in its new position, the result will be zero. Therefore, temporal differences are not very sensitive to background changes, but they can often pose problems when detecting inner parts of moving objects. Therefore, the object can appear to be hollow in the temporal difference image. The temporal difference between frames is calculated by taking the difference of a neighborhood of M × M pixels and then the absolute value of the result:

$$G_2(m,n,k) = \frac{1}{M^2} \left| \sum_{p=1}^{M-1} \sum_{q=1}^{M-1} F(p,q,k) - B(p,q,k-1) \right|. \tag{13.2}$$

The image G_2 is calculated for both the current frame together with the previous frame as well as the next frame together with the current frame. Figure 13.3 shows the application of the temporal difference methodology to the selected image sequence. Some postprocessing has been employed in this case, which removes the spurious noise with median filtering and enhances the connected pixels with a combination of dilation and erosion operations.

Although the target has been missed in frame 3 owing to postprocessing, the localization in the adjacent frames is sufficient information for the machine vision system to follow the target.

13.4.1 Gradient Difference

There are always small changes in illumination and small movements between the frames. These changes affect the subtractions and cause artifacts, even though they are hard for the human eye to see. The gradient of an image, on the other hand, is not so sensitive to time-varying illumination changes. Performing a gradient difference on our image sequences

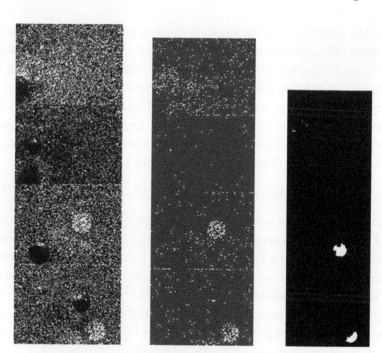

(a) (b) (c)

FIGURE 13.3 Object detection with temporal frame difference for the source image of Figure 13.2 (a). (a) Difference image in grayscale, (b) difference image after being converted to a binary image, (c) images after postprocessing.

will therefore be important in removing artifacts of the type mentioned earlier. The gradient of a frame is generated by taking the gradient in both x and y directions, using the Sobel operator. The current frame is then subtracted from the reference background for each direction separately.

13.5 CORRELATION-BASED TRACKING

This method is primarily a shape-based technique and can work equally well with both RGB as well as grayscale images. Cross-correlation is a standard method of estimating the degree to which two images are correlated. Consider two images $x(m,n)$ and $y(m,n)$, where $n = 0, 1, 2, ..., N-1$. The cross-correlation r at delay δ is defined as

$$\rho_{xy}(m,n,p,q) = E[x(m,n)y(m+p,n+q)] = \sum_{p=0}^{M}\sum_{q=0}^{N} x(m,n)y(m+p,n+q).$$

(13.3)

Due to the complexity of the implementation, a better approach that utilizes some interesting properties of the Fourier transform is presented here. If one of the images is symmetric, then flipping it columnwise will not make any difference and, hence, the correlation and convolution can be treated in the same way. Because convolution in the spatial domain becomes pixel-by-pixel multiplication in the frequency domain, which is a much faster process than convolution, the cross-correlation can, therefore, be applied using Fourier transforms of the two images. Hence:

$$R_{xy} = X(u,v)Y(u,v) \quad \Rightarrow \quad \rho_{xy} = \Im^{-1}R_{xy}. \qquad (13.4)$$

Now, consider the case when one of the foregoing two images is the target object in the source image, which itself is the second image. Wherever there is a perfect match between the source image and the target image, the cross-correlation is also very large. The resulting correlation image is then thresholded to isolate the location points.

Figure 13.4 shows an interesting example of the application of such a technique in which the goal is to count the number of trees present in the

(a)

(b)

FIGURE 13.4 Using cross-correlation-based target tracking. (a) Original scene with trees and target is highlighted with dotted rectangle, (b) result of correlation highlighting the peaks.

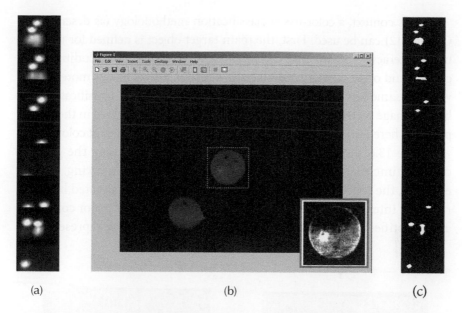

(a)	(b)	(c)

FIGURE 13.5 Object detection with cross-correlation method. (a) Grayscale image with target highlighted as dotted rectangle and is also reproduced in the inset, (b) resulting grayscale images after correlation, (c) images after postprocessing.

scene. In this case, a particular treetop image is selected as the object, and is located in the whole image using the convolution method. The resulting image is then thresholded for the points of maximum correlation, which ultimately gives the tree count.

The same technique is applied to the sample images of Figure 13.2(a), and the results are shown in Figure 13.5. As can be seen in Figure 13.5, the exact location is slightly shifted from the center of the target image due to convolutional drifts.

13.6 COLOR-BASED TRACKING

Tracking objects based on color is one of the quickest and easiest methods for tracking an object from one image frame to the next. The speed of this technique makes it very attractive for near real-time applications, but due to its simplicity, many issues exist that can cause the tracking to fail.

In this context, a color-based classification methodology (as described in Chapter 12) can be used. First, the main target object is defined for its color using interactive or automated methods. All other colors, including background and other unwanted artifacts, are recognized as one or more classes with the same representative pseudocolor. The result of the classification is a binary image with the targeted object displayed in one color (red in the example given here), whereas everything else is displayed in a different color.

Figure 13.6 shows the procedure for class selection using the graphical ROI function. Figure 13.7 shows the results of such tracking. In this example, the red balloon is first identified, with its color selected by some region of interest (ROI) selection method. Similarly, other major colors are also identified in the same fashion and assigned the same representative

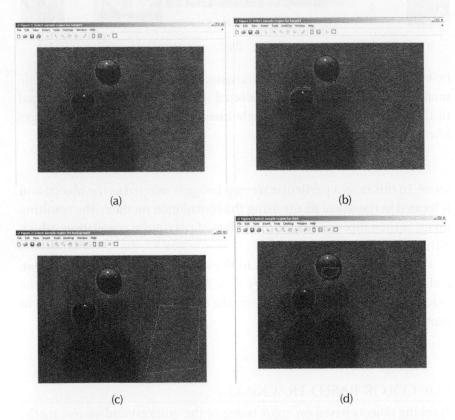

(a) (b)

(c) (d)

FIGURE 13.6 (See color insert following Page 204.) Declaring various classes for color-based tracking. (a) Target 1 is highlighted for red color, (b) target 2 is highlighted for blue color, (c) background is highlighted, (d) dark areas are highlighted.

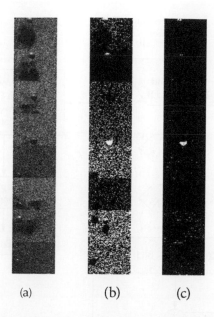

(a)   (b)   (c)

FIGURE 13.7 Object detection with color-based classification method. (a) RGB images with background removed, (b) resulting binary image with identified classes, (c) images after postprocessing.

class color, so that all the nontarget artifacts and background will be represented by the same color and will appear as a complete class that is other than the actual target class.

The results with color processing are obtained faster and can be embedded in a machine vision system at the camera level. The target color can be filtered at the acquisition level where all other colors can be blocked so that only the selected color will reach the sensors. This makes the tracking very easy for the machine vision system because most of the error-producing elements have already been removed.

13.7 ALGORITHMIC ACCOUNT

The four tracking algorithms presented in the previous sections can be represented as flow charts for coding in any environment. Figures 13.8–13.11 represent these algorithms.

13.8 MATLAB CODE

In this section, the MATLAB code used to generate the images in Figures 13.2, 13.3, 13.5, 13.6, and 13.7 is presented. We begin with the background subtraction method of Figure 13.2: A file track1.m is needed

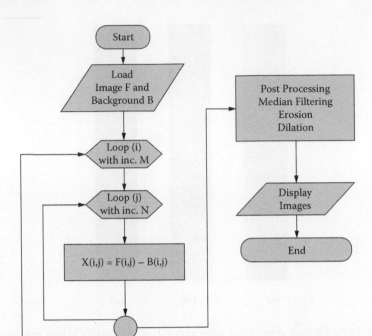

FIGURE 13.8 Algorithmic logic for background subtraction method.

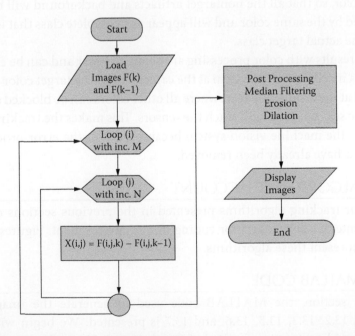

FIGURE 13.9 Algorithmic logic for temporal frame subtraction method.

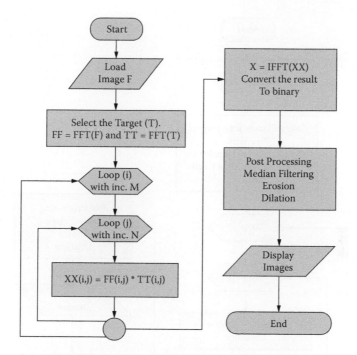

FIGURE 13.10   Algorithmic logic for correlation method.

first to generate the necessary image files. However, the test images are also stored in binary files, namely, x1.mat, x2.mat, x3.mat, and x4.mat, corresponding to single red object, single blue object, two objects, and three objects, respectively.

```
% First Run track1.m to get the mat files
% This file uses background subtraction only
clear all
close all

load x1
bk = imread('p00.jpg') ;
XG = [] ;
XB = [] ;
ro = 480 ; co = 640 ;
for i = 1 : length(X1)/ro
    x = X1((i-1)*ro+1:i*ro,:,:) ;
    x = bk - x ;
    xg = rgb2gray(x) ;
    xg(ro:-1:ro-2, 1:co) = 255 ;
```

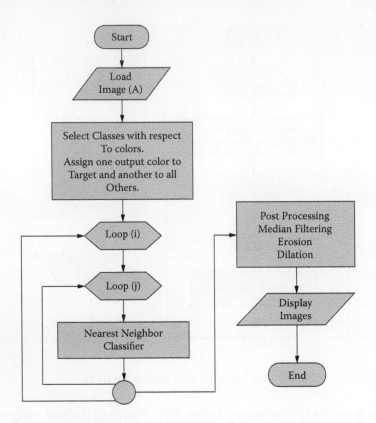

FIGURE 13.11 Algorithmic logic for color-based classification method.

```
        xb = im2bw(xg,0.1) ;
        XG = [ XG ; xg ] ;
        XB = [ XB ; xb ] ;
end

figure ; imshow(X1)
figure ; imshow(XG)
figure ; imshow(XB)
```

For the temporal frame subtraction strategy of Figure 13.3, the following code was used:

```
% First Run track1.m to get the mat files
% This file uses inter-frame difference
clear all
close all

load x1
```

```
bk = imread('p00.jpg') ;
XG = [] ;
XB = [] ;
Z1G = [] ;
Z1B = [] ;

ro = 480 ; co = 640 ;

for i = 1 : length(X1)/ro
    x = X1((i-1)*ro+1:i*ro,:,:) ;
    x = bk - x ;
    xg = rgb2gray(x) ;
    xg(ro:-1:ro-2, 1:co) = 255 ;
    xb = im2bw(xg,0.1) ;
    XG = [ XG ; xg ] ;
    XB = [ XB ; xb ] ;
  if i > 1
    y = X1((i-2)*ro+1:(i-1)*ro,:,:) ;
    y = y - bk ;
    yg = rgb2gray(y) ;
    z1g = yg - xg ;
    z1g = yg - z1g ;
    z1b = im2bw(z1g,0.01) ;
    Z1G = [ Z1G ; z1g ] ;
    Z1B = [ Z1B ; z1b ] ;
  end
end
figure ; imshow(X1)
figure ; imshow(Z1G*100)
figure ; imshow(Z1B)

% Post-processing
Z2B = medfilt2(Z1B,[3 3]);
y = bwmorph(Z2B,'dilate',5) ;
Z2B = medfilt2(y,[15 15]);
y = bwmorph(Z2B,'erode',7) ;
figure ; imshow(y)
```

The correlation method was implemented using the Fourier transformation method, and the required code is as follows:

```
% First Run track1.m to get the mat files
% This file uses correlation of target in grayscale
```

```
clear all
close all

load x3
X1 = X3 ;
bk = imread('p00.jpg') ;
XG = [] ;
Z1G = [] ;
Z1B = [] ;

ro = 480 ; co = 640 ;
p = 3 ; % image # (select 4 for single object (x1.mat)
or 3 for two object (x3.mat) case
  xm = X1((p-1)*ro+1:p*ro,:,:) ;
  xm = bk - xm ;
  xmg = rgb2gray(xm) ;
  figure ; imshow(xmg) ;
  rect = getrect(gca) ;
  xmm = xm(rect(2):rect(2)+rect(3)-
1,rect(1):rect(1)+rect(4)-1);
  XMM = fft2(double(xmm),1024,1024) ;
   figure ; imshow(xmm*10)

for i = 1 : length(X1)/ro
    x = X1((i-1)*ro+1:i*ro,:,:) ;
    x = bk - x ;
    xg = rgb2gray(x) ;
    y = xg ;
    y(ro:-1:ro-2, 1:co) = 255 ;
    XG = [ XG ; y ] ;
    XMG = fft2(double(xg),1024,1024) ;
    Y = XMG .* XMM ;
    y = ifft2(Y) ;
    y = y(1:ro,1:co);
    x = y*0.0000004 ;
    z=im2bw(x);
    z = bwmorph(z,'erode',19) ;
    Z1G = [ Z1G ; x ] ;
    Z1B = [ Z1B ; z ] ;
end

figure ; imshow(Z1G)
figure ; imshow(Z1B)
```

Finally, the color-based classification code:

```
% First Run track1.m to get the mat files
% This file uses color classification method
clear all
close all

load x3
X1 = X3 ;
bk = imread('p00.jpg') ;
X = [] ;
XC = [] ;
Z1G = [] ;
Z1B = [] ;
S = 10 ;

ro = 480 ; co = 640 ;
p = 2 ; % image #
x = X1((p-1)*ro+1:p*ro,:,:) ;
x = x - bk ;
x = x * S ;
figure ; imshow(x) ;

[r,c,s] = size(x) ;

% Initialize storage for each sample region.
classes = { 'target1','target2','background', 'dark' };
nClasses = length(classes);
sample_regions = false([r c nClasses]);

% Select each sample region.
f = figure;
for count = 1:nClasses
    set(f, 'name', ['Select sample region for '
classes{count}] );
    sample_regions(:,:,count) = roipoly(x);
end
close(f);

for i = 1 : length(X1)/ro
    x = X1((i-1)*ro+1:i*ro,:,:) ;
    % Convert the RGB image into an L*a*b* image
    x = x - bk ;
```

```
    x = x * S ;
    X = [ X ; x ] ;
    cform = makecform('srgb2lab');
    lab_x = applycform(x,cform);

 % Calculate the mean 'a*' and 'b*' value for each ROI
area
 a = lab_x(:,:,2);
 b = lab_x(:,:,3);
 color_markers = repmat(0, [nClasses, 2]);
 for count = 1:nClasses

 %      color_markers(count,1) =
median(a(sample_regions(:,:,count)));
 %      color_markers(count,2) =
median(b(sample_regions(:,:,count)));
      color_markers(count,1) =
mean2(a(sample_regions(:,:,count)));
      color_markers(count,2) =
mean2(b(sample_regions(:,:,count)));
 end

 %% Classify Each Pixel Using the Nearest Neighbor Rule
 color_labels = 0:nClasses-1;
 a = double(a);
 b = double(b);
 distance = repmat(0,[size(a), nClasses]);

 % Perform classification
 for count = 1:nClasses
     distance(:,:,count) = ( (a -
color_markers(count,1)).^2 + ...
            (b - color_markers(count,2)).^2 ).^0.5;
 end

 [value, label] = min(distance,[],3);
 label = color_labels(label);
 colors = [ 255 255 255 ; 0 0 0 ; 0 0 0 ; 0 0 0 ] ;
 y = zeros(size(x)) ;
 l = double(label)+1 ;
 for m = 1 : r
     for n = 1 : c
         y(m,n,:) = colors(l(m,n),:) ;
     end
```

```
end

figure ; imshow(y)
XC = [ XC ; y ] ;
z=im2bw(y) ;
z(ro:-1:ro-4, 1:co) = 255 ;
w = bwmorph(z,'erode',3) ;
Z1G = [ Z1G ; w ] ;
Z1B = [ Z1B ; z ] ;
end

figure ; imshow(X)
figure ; imshow(XC)
figure ; imshow(Z1G)
figure ; imshow(Z1B)
```

13.9 SUMMARY

- Target tracking is a very important technique used in a number of industrial and defense-related applications.

- The basic idea is to enhance the changes occurring in subsequent frames in a sequence of images or video.

- The changes usually refer to the objects of interest, including a particular target object.

- Many techniques have been developed for various application domains and environments.

- However, the fundamental technique of removing the static portions of an image such as background, etc., is part of all of these techniques.

- Four such fundamental techniques are presented in this chapter to track a moving artifact in an image sequence with a fixed background.

- The first method is used only in a scenario in which just one object is under consideration and can be extracted simply by subtracting the static background image.

- In the second method, in addition to background removal, temporal frames are also subtracted to track any moving object in the time interval between the frames.

- The third approach is only related to the shape of the object to be tracked and deals with correlation of the target image. Here, the target's subimage is correlated with the new frame's image, and it gives a high correlation value when the object is matched anywhere in the image.

- This method is usually used with shape-based target tracking irrespective of the grayscale values or colors.

- The fourth method is based on detecting a specific color to be classified.

- The classification is repeated in each frame and, hence, the same colored object is tracked.

- A lot of postprocessing is usually needed to precisely isolate the targets from the clutter of artifacts resulting from the difference or correlation operations.

- The postprocessing usually includes median filtering and morphological operations.

13.10 EXERCISES

1. Can method 1 (background subtraction) work for more than one moving object? If not, suggest a modification in subtraction sequence or other strategy to increase the efficiency of this method.

2. Can method 2 (temporal frame difference) work for more than one moving object? If not, suggest a modification in subtraction sequence or other strategy to increase the efficiency of this method.

3. Repeat method 3 (the correlation method) for more than one object in motion. Comment on your observations.

4. Can method 3 be used for classifying objects of similar shape but different colors/shades? Explain why or why not.

5. Repeat the color-based classification method with a grayscale image. Comment on your observations.

CHAPTER 14

Face Recognition

14.1 INTRODUCTION

F ACE RECOGNITION REFERS TO the technique of identifying a given face image within an image database. It is different from the operation of face detection, which essentially means to find the face of any human subject in a given image without knowing who the subject is. Face recognition has received much attention in recent years due to its many applications in different fields such as law enforcement, security applications, or video indexing. Face recognition is a very challenging problem and to date, there is no technique that provides a robust solution in all situations and various applications that face recognition may encounter.

Face recognition can be divided into two basic applications: identification and verification. In the identification problem, the face to be recognized is unknown and is matched against faces in a database containing known individuals. In the verification problem, the system confirms or rejects the claimed identity of the input face. Again, the problem is still an open area of research, and a single technique that can outclass every other technique is still awaited.

14.2 FACE RECOGNITION APPROACHES

Face recognition approaches to still images can be broadly grouped into two approaches:

- Geometric

- Template-matching techniques

In the first case, geometric characteristics of faces to be matched, such as distances between different facial features, are compared. This technique provides limited results, although it has been used extensively in the past. In the second case, face images represented as a two-dimensional array of pixel intensity values are compared to a single or several templates representing the whole face. More successful template-matching approaches use principal components analysis (PCA) or linear discriminant analysis (LDA) to perform dimensionality reduction, achieving good performance at a reasonable computational complexity/time. Other template-matching methods use neural network classification and deformable templates, such as elastic graph matching (EGM).

The appearance of the human face image has potentially very large intrasubject variations due to 3D head pose:

- Illumination (including indoor/outdoor)
- Facial expression
- Face angle or position
- Occlusion due to other objects or accessories (e.g., sunglasses, scarf, etc.)
- Facial hair
- Aging

A number of face recognition algorithms, along with their modifications, have been developed during the past several decades (see Figure 14.1). As can be seen from the figure, each domain has its own techniques and algorithms to perform the task. By no means is this an exhaustive representation of all the existing techniques; newer and better techniques keep emerging. However, this figure gives an overall idea of where the focus should be for a particular group with a specific interest/expertise.

In this chapter, the focus is on the use of PCA with statistical distance measurements to distinguish faces from one another. As can be sensed from these statements, all this requires a lot of matrix manipulation, and this is discussed next.

14.3 VECTOR REPRESENTATION OF IMAGES

Image data can be represented as vectors, that is, as points in a high-dimensional vector space. For example, a $p \times q$ 2D image can be mapped to a vector $x \in \Re^{pq}$ by lexicographic ordering of the pixel elements (such

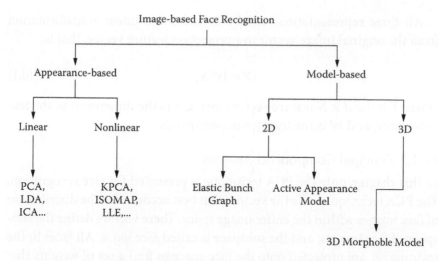

FIGURE 14.1 Various techniques used in face recognition.

as by concatenating each row or column of the image). Despite this high-dimensional embedding, the natural constraints of the physical world (and the imaging process) dictate that the data will, in fact, lie in a lower dimensional (though possibly disjoint) manifold. The primary goal of the subspace analysis is to identify, represent, and parameterize this manifold in accordance with some optimality criteria.

Let $X = \{x_1 \quad x_2 \quad \cdots \quad x_n\}$ represent an $n \times N$ data matrix, where each x_i is a face vector of dimension n, concatenated from a $p \times q$ face image, where $n = p \times q$. Here, n represents the total number of pixels in the face image, and N is the number of different face images in the training set. The mean vector of the training images, $\mu = \Sigma_1^N x_i$, is subtracted from each image vector.

14.3.1 Linear (Subspace) Analysis

Three classical linear appearance-based classifiers, PCA, ICA, and LDA, are commonly used for face recognition, once the image is represented in its vector space. Each classifier has its own representation (basis vectors) of a high-dimensional face vector space based on different statistical viewpoints. By projecting the face vector to the basis vectors, the projection coefficients are used as the feature representation of each face image. The matching score between the test face image and the training prototype is calculated (e.g., as the cosine value of the angle) between their coefficients vectors. The larger the matching score, the better the match.

All three representations can be considered a linear transformation from the original image vector to a projection feature vector, that is,

$$Y = W^T X, \tag{14.1}$$

where Y is the $d \times N$ feature vector matrix, d is the dimension of the feature vector, and W is the transformation matrix.

14.3.2 Principal Components Analysis

In this chapter, only the PCA technique is presented for face recognition. The PCA technique finds the vectors that best account for the distribution of face images within the entire image space. These vectors define the subspace of face images, and the subspace is called *face space*. All faces in the training set are projected onto the face space to find a set of weights that describes the contribution of each vector in the face space. Identifying a test image requires the projection of the test image onto the face space to obtain the corresponding set of weights. By comparing the weights of the test image with the set of weights of the faces in the training set, the face in the test image can be identified.

The key procedure in PCA is based on the Karhumen–Loeve transformation. If the image elements are considered to be random variables, the image may be seen as a sample of a stochastic process. The PCA basis vectors are defined as the eigenvectors of the scatter matrix S_T,

$$S_T = \sum_{i=1}^{N} (x_i - \mu)(x_i - \mu)^T. \tag{14.2}$$

The transformation matrix W_{PCA} is composed of the eigenvectors corresponding to the d largest eigen values.

After applying the projection, the input vector (face) in an n-dimensional space is reduced to a feature vector in a d-dimensional subspace. For most applications, the eigenvectors corresponding to very small eigen values are considered to be noise, and are not taken into account during identification. Several extensions of PCA are developed, such as modular eigenspaces and probabilistic subspaces.

14.3.3 Databases and Performance Evaluation

A number of face databases have been collected for different face recognition tasks. Table 14.1 lists a selection of those available in the public

TABLE 14.1   Selected Face Databases Available in the Public Domain

Face database	No. of subjects	No. of images	Types of variations
ORL	40	400	P, E
Yale	15	165	I, E
AR	>120	>3000	I, E, O, T
MIT	16	432	I, P, S
UMIST	20	564	P
CMU PIE	68	41368	P, I, E
XM2VTS	295	>3000	N/A
FERET	>1000	>10000	P, I, E, T

Note: Types of variations are abbreviated as follows: E: expression; I: illumination; O: occlusion; P: pose; S: scale; T: time interval [images of the same subject are taken between a short period (e.g., a couple of days) or a long period (e.g., years)].

domain. The AR database contains occlusions due to eye glasses and scarf. The CMU PIE database is a collection with well-constrained poses, illumination, and expression. The FERET and XM2VTS databases are the two most comprehensive databases, which can be used as a benchmark for detailed testing or comparison. The XM2VTS is especially designed for multimodal biometrics, including audio and video cues. To keep facial recognition technology evaluation abreast of state-of-the-art advances, the Face Recognition Vendor Test (FRVT) followed the original FERET, and was conducted in the year 2000 and 2002 (namely, FRVT2000 and FRVT2002). The database used in FRVT was significantly extended between 2000 and 2002, including more than 120,000 face images from more than 30,000 subjects. More facial appearance variations were also considered in FRVT, such as indoor/outdoor difference.

Obviously, in order to test the system, some faces are required. The following example described here has been executed on some of the faces provided by the ORL face database (as shown in Figure 14.2).

A set of randomly selected faces along with their class tags were used for training. In considering the training part of the face database, a preprocessing step is needed in which the faces are in some way normalized. In fact, they are approximately centered, all at the same scale, with a roughly equivalent background for each picture. In the testing set, the faces are not normalized at all. They are not centered, exhibit a wide variety of scales, and include greater variation in background. Facial expressions are dissimilar (sometimes smiling, sometimes sad, etc.), hair is not necessarily combed, and people often

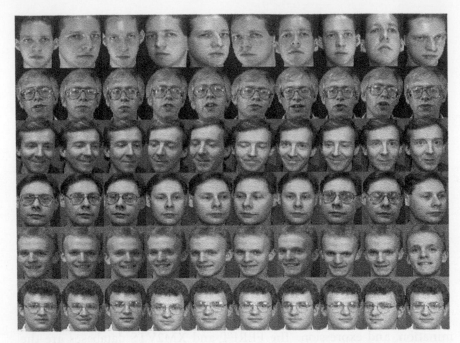

FIGURE 14.2 Part of the testing set from the ORL face database used in this chapter.

wear glasses. As a technical aside, it should be noted that each picture's dimension is 112 × 92 pixels and each pixel is coded on 8 bits (256 gray levels).

14.4 PROCESS DETAILS

The process by which the recognition results are obtained must be well understood. As has been said, PCA computes the basis of a space, which is represented by its training vectors. The basis vectors computed by PCA are in the direction of the largest variance of the training vectors. These basis vectors are computed by solution of an eigen problem and, as such, the basis vectors are eigenvectors. Figure 14.3 shows the same figure as in Figure 14.2 but in its eigenface form.

These eigenvectors are defined in the image space. They can be viewed as images and, indeed, look like faces. Hence, they are usually referred to as eigenfaces. The first eigenface is the average face, and the rest of the eigenfaces represent variations from this average face. The first eigenface is a good face filter: each face multiplied pixel by pixel (inner product) with this average face yields a number close to one; with nonface images, the

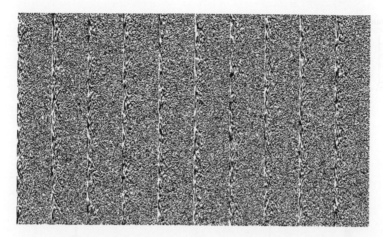

FIGURE 14.3 Each image of Figure 14.2 in its eigenspace representation.

inner product is much less than one. The direction of the largest variation (from the average) of the training vectors is described by the second eigenface. The direction of the second-largest variation (from the average) of the training vectors is described by the third eigenface, and so on.

Each eigenface can be viewed as a feature. When a particular face is projected onto the face space, its vector (made up of its weight values with respect to each eigenface) into the face space describes the importance of each of those features in the face. Figure 14.4 shows schematically what PCA does. It takes the training faces as input and yields the eigenfaces as output. Obviously, the first step of any experiment is to compute the eigenfaces. Once this is done, the identification or categorization process can begin.

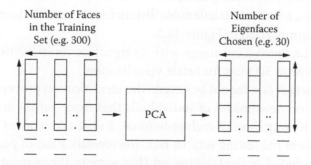

FIGURE 14.4 Eigenface generation process.

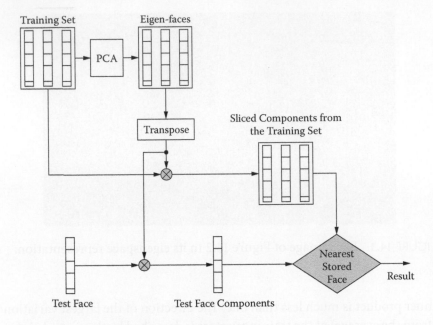

FIGURE 14.5   Face identification process.

Once the eigenfaces have been computed, the face space has to be populated with known faces. Usually, these faces are taken from the training set. Each known face is transformed into the face space, and its components stored in memory.

At this stage, the identification process can begin. An unknown face is presented to the system. The system projects it onto the face space and computes its distance from all the stored faces. The face is identified as being the same individual's as the face that is nearest to it in face space. There are several methods of computing the distance between multidimensional vectors. Here, a form of Mahalanobis distance is chosen. The identification process is summarized in Figure 14.5.

Figure 14.6 shows a test image with its eigenface components shown as images as well as 3D plots for better visualization.

An important fact should be noted. The identification process has been tested only on new images of individuals that made up the training set (out of the 10 mug shots of each individual, 2 were taken out of the training set to form the testing set). In fact, the identification of persons who were not included in the training set (but were in the populating set) is often very poor.

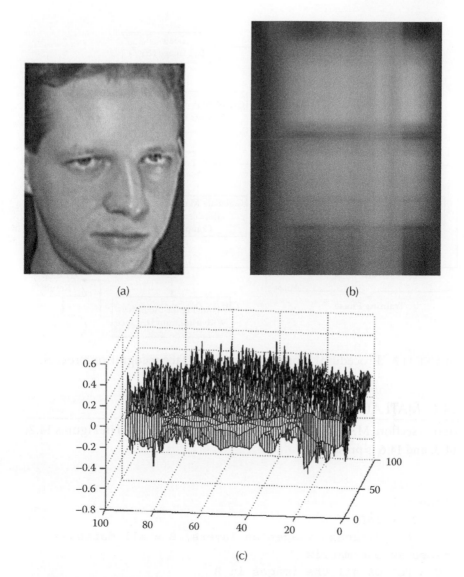

(a)   (b)

(c)

FIGURE 14.6 Results from a test run. (a) Test image; (b) PCA components of (a); (c) 3D view of (b).

14.5 ALGORITHMIC ACCOUNT

A procedural flow chart for the preceding technique is shown in Figure 14.7. The generic blocks represent the main processing parts; however, the details of each part may be subject to the actual implementation environment and should be modified accordingly.

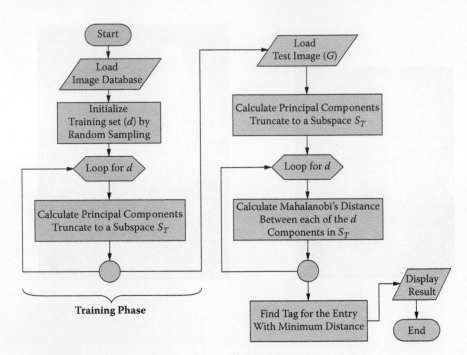

FIGURE 14.7   Procedural representation of the face recognition process.

14.6 MATLAB® CODE

In this section, MATLAB code used to generate the images in Figures 14.2, 14.3, and 14.6 is presented:

```
clear all
close all
[A,B,C] = ldf ;
% A = All database images as layers, B = all database
images as one matrix
% C = PCA of all the images in B

N = 6 ; % No. of People
M = 10 ; % No. of poses per person

[r,c] = size(B) ;
n = r / N ; % # rows in each image
m = c / M ; % # cols in each image

T = B(:,1:m) ; % Training images
Tp = C(:,1:m) ; % PCA of training images
G = ones(r,1) ;
```

```
for i = 1 : N
    G((i-1)*n+1:i*n,1) = i ;
end

tc = floor(rand(1)*20+1) ;
tc = 15 ;
pd = princomp(A(:,:,tc)) ;
%[c1,err,post,logl,str] = classify(pd,T,Tp,'mahalnobis');
for i = 1 : N
    t1 = Tp((i-1)*92+1:i*92,:) ; % Selecting each image PCA
    mhh = mahal(pd',t1') ; % Calculating the Mahalanobis
distance
    mh(i) = min(mhh) ;
end

[p,q] = min(mh) ;
% q will give the class # for the test image
[tc q]
```

14.7 SUMMARY

- Human face recognition refers to the process of identifying an individual from a known set of people.

- Such sets have been very carefully assembled and have been made available as databanks or repositories, mostly without any charge.

- The main challenge is the fact that any test image when compared with the images in this database may not have the same orientation, illumination, expressions, or other features as those present in the database.

- As such, one-to-one matching is not possible.

- Usually, some form of mapping is employed to convert the image into some representative components that are fairly independent of the aforementioned differences.

- PCA is one such technique that can decompose a face image into its eigen components, which can be used for classification.

- Once such components are known for all the images in the database, any test image's components are compared with the existing components.

- The comparison is usually made in the form of distance calculation of some type.

- The class in the database to which the distance is found to be least is then declared to be the recognition information for the test image.

14.8 EXERCISES

1. Modify the given code for Euler's distance instead of Mahalanobis distance. Comment on the performance in comparison with the one given.

2. Other decompositions can also be tried instead of PCA. For instance, use singular value decomposition (svd) to get the eigen components, and repeat the given code. Compare and comment on the results.

3. Repeat Q #2 for Euler's distance.

CHAPTER 15

Soft Computing in Image Processing

15.1 INTRODUCTION

F OR THE PAST THREE decades, there have been two major groups of researchers in the field of algorithmic development for real systems; the first group believes that such development can only be done by using conventional mathematical and probabilistic techniques, whereas the second group emphasizes that there are other methods of applying mathematical knowledge that need not be that restrictive in terms of boundaries and samples. Soft computing differs from conventional (hard) computing in that, unlike hard computing, it is tolerant of imprecision, uncertainty, partial truth, and approximation. In effect, the role model for soft computing is the human mind. Certainly, the way our brain works is different from the way a microprocessor works because our brain can get an intuitive "feel" of things rather than exact measured values, and its estimates are perceptive rather than numeric. Hence, the guiding principle of soft computing is this: Exploit the tolerance for imprecision, uncertainty, partial truth, and approximation to achieve tractability, robustness, and low solution cost. The basic ideas underlying soft computing in its current incarnation have links to many earlier influences, with Zadeh's 1965 paper on fuzzy sets holding a pioneering position. In fact, for many years, soft computing was being referred to as fuzzy logic as well. However, the set has grown since, and now the principal constituents of soft computing are

- Fuzzy systems
- Neural networks
  - Precetron-based
  - Radial basis functions
  - Self-organizing maps
- Evolutionary computation
  - Evolutionary algorithms
  - Genetic algorithms
- Harmony search
- Swarm intelligence
- Machine learning
- Chaos theory

What is important to note is that soft computing is not an amalgam of various techniques; rather, it is a partnership in which each of the components contributes a distinct methodology to address problems in its domain. This notion has an important consequence: in many cases, a problem can be solved most effectively by using a combination of the constituent techniques rather than exclusively by any single technique. A striking example of a particularly effective combination is what has come to be known as *neuro-fuzzy systems*. Such systems are becoming increasingly visible as consumer products, ranging from air conditioners and washing machines to photocopiers and camcorders.

Soft computing attempts to study, model, and analyze very complex phenomena, those for which more conventional methods have not yielded low-cost, analytic, and complete solutions. Earlier computational approaches could model and precisely analyze only relatively simple systems. More complex systems arising in biology, medicine, the humanities, management sciences, and similar fields often remained intractable to conventional mathematical and analytical methods.

15.2 FUZZY LOGIC IN IMAGE PROCESSING

This chapter presents only fuzzy logic and its applications in the image processing domain because it would be impossible to cover all the soft computing constituent technologies in this book. One added advantage of working with fuzzy logic is its simplicity and the ease with which the underlying concepts can be understood. This will be explained in the following sections using two examples related to image classification and image filtering.

Fuzzy set theory is the extension of conventional (crisp) set theory. It handles the concept of partial truth [truth values between 1 (completely true) and 0 (completely false)]. It was introduced by Prof. Lotfi A. Zadeh of the University of California at Berkeley in 1965 as a mean of modeling the vagueness and ambiguity in complex systems. Since then, it has been used in increasingly diversified applications covering almost every discipline in science and technology.

The application of fuzzy logic to image processing can be explained in light of the actual working mechanism of the logic itself, as shown in Figure 15.1. The system shown in Figure 15.1 requires expert knowledge related to the image and must undergo necessary processing before it can be incorporated as an embedded part of the rule base or other functionalities of fuzzy logic. For instance, knowledge of changing light intensity in an irregularly illuminated image, imparted to the system by the user, will

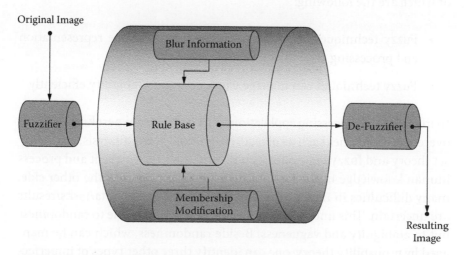

FIGURE 15.1 Functional block diagram for a fuzzy image processing system.

be used to modify the membership function in the RUN mode so that it may be adjusted/adapted to the desired thresholds. This can be utilized as an adaptive edge detector.

Similarly, for uneven blurring, expert knowledge will identify the type of blur on the fly, and membership functions as well as the rule base can be modified accordingly. Yet another example could be a binary image converter. For instance, one may want to define a set of gray levels that share the property "dark." In classical set theory, a threshold has to be determined, say, the gray level 100. All gray levels between 0 and 100 are elements of this set, whereas the others do not belong to the set [Figure 15.2(a)]. But darkness is a matter of degree. So, a fuzzy set can model this property much better. To define this set, one also needs two thresholds, say, gray levels 50 and 150. All gray levels that are less than 50 are full members of the set, whereas all gray levels that are greater than 150 are not the members of the set. The gray levels between 50 and 150, however, have a partial membership in the set as shown in Figure 15.2(b). Figure 15.2(c) depicts this idea with respect to the functionality diagram for fuzzy logic.

15.2.1 Why Fuzzy Image Processing?

Before proceeding further to actual examples, it would be appropriate to answer a legitimate question at this point: "Why should one use fuzzy techniques in image processing?" There are many reasons, the most important of which are the following:

- Fuzzy techniques are powerful tools for knowledge representation and processing.

- Fuzzy techniques can manage vagueness and ambiguity efficiently.

In many image processing applications, one has to use expert knowledge to overcome difficulties such as object recognition, scene analysis, etc. Fuzzy set theory and fuzzy logic offer us powerful tools to represent and process human knowledge in the form of fuzzy if–then rules. On the other side, many difficulties in image processing arise because the data/tasks/results are uncertain. This uncertainty, however, is not always due to randomness but to ambiguity and vagueness. Beside randomness, which can be managed by probability theory, one can identify three other types of imperfection in routine image processing:

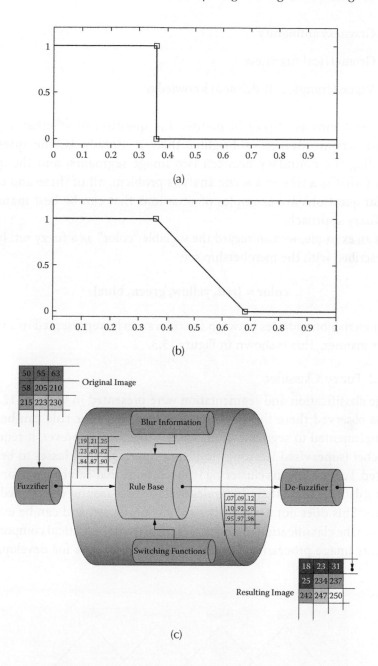

FIGURE 15.2 Explanation of fuzzy logic application. (a) Crisp membership function, (b) actual fuzzy membership function, (c) sample application of blur removal.

- Grayness ambiguity

- Geometrical fuzziness

- Vague (complex/ill-defined) knowledge

These problems are fuzzy in nature. The question of whether a pixel should become darker or brighter than it already is, the question regarding the boundary between two image segments, and the question of what is a tree in a scene analysis problem, all of these and other similar questions are examples of situations that can be best managed by a fuzzy approach.

As an example, we can regard the variable "color" as a fuzzy set. It can be described with the membership set:

color = {red, yellow, green, blue}

The noncrisp boundaries between the colors can be represented in a much better manner. This is shown in Figure 15.3.

15.2.2 Fuzzy Classifier

Image classification and segmentation were presented in Chapter 12, and it was observed there that the simple nearest neighbor rule can be easily implemented to segment similar pixels together. However, it required a teacher (supervised learning) to identify parts of the classes to be segmented. Fuzzy c-means clustering is a technique that can be used for classifying data based on their dynamics of clustering around certain deduced centers. This does not necessarily need a training set, and can be used to perform the classification on images. The different theoretical components of fuzzy image processing provide diverse possibilities for development

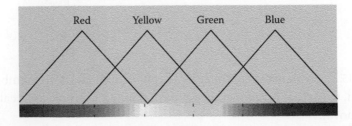

FIGURE 15.3 Fuzzy membership representing color space.

of new segmentation techniques. The following list is a brief overview of different fuzzy approaches to image segmentation:

Fuzzy clustering algorithms: Fuzzy clustering is the oldest fuzzy approach to image segmentation. Algorithms such as fuzzy c-means and probabilistic c-means can be used to build clusters (segments). The class membership of pixels can be interpreted as similarity or compatibility with an ideal object or a certain property.

Fuzzy rule-based approach: If image features are interpreted as linguistic variables, one can use fuzzy if–then rules to segment the image into different regions. A simple fuzzy segmentation rule may appear as follows:

IF the pixel is *dark*

AND its neighborhood is also *dark* **AND** *homogeneous*

THEN it *belongs* to the background.

Measures of fuzziness and image information: Measures of fuzziness (e.g., fuzzy entropy) and image information (e.g., fuzzy divergence) can be also used in segmentation and thresholding tasks that are commonly deployed in image processing algorithms.

Fuzzy geometry: Fuzzy geometrical measures such as fuzzy compactness and index of area coverage can be used to measure the geometrical fuzziness of different regions of an image. The optimization of these measures (e.g., minimization of fuzzy compactness regarding the cross-over point of membership function) can be applied to make fuzzy and/or crisp pixel classifications.

15.2.2.1 Fuzzy C-Means Clustering

In this book, we have utilized the well-known fuzzy c-means clustering (FCMC) algorithm to reduce errors in GPS readings by mapping proximity of data points from GPS into its cluster centers. Furthermore, a dual-GPS receiving system is utilized that helps in reducing the error even further, because the actual position coordinate will now be the average of the most prominent cluster centers from the two sensors along the vehicle's axis. The hardware was interfaced with LabView 7.1 for all the sensors, GPS

receivers, and control actuators. FCMC was implemented in MATLAB®
6.5 and was incorporated with the LabView files.

FCMC is a data-clustering technique in which each data point belongs to a
cluster to some degree that is specified by a membership grade. This technique
was originally introduced by Jim Bezdek in 1981 as an improvement on earlier
clustering methods. It provides a method of grouping data points that popu-
late some multidimensional space into a specific number of different clusters.

A fuzzy c-partition of X is defined by a $(c*n)$ matrix $U = [u_{ik}]$, where $u_{ik} = u_i(x_k)$ is the degree of membership of x_k in the ith cluster u_i and $\{ u_i : x \rightarrow [0,1] \}$. The following properties must be true:

$$u_{ik} \in [0,1], \qquad \forall i,k,$$

$$\sum u_{ik} = 1, \qquad \forall k, \tag{15.1}$$

$$0 < \sum u_{ik} < n, \qquad \forall i.$$

Because of this distribution of membership among the c fuzzy clusters, fuzzy
c-partition provides much more information about the structure in each data
cluster than does hard c-partition. Then, to solve the following problem:

$$minimize \; J_m(U,V) = \sum_i \sum_k u_{ik}^m d^2(x_k, V_i), \tag{15.2}$$

with respect to $U = [u_{ik}]$, a fuzzy c-partition of n unlabeled data sets $X = \{x_1, ..., x_n\}$, and to V, a set of c fuzzy cluster centers $V = (V_1, ..., V_c)$, the parameter
$m > 1$ is used as the fuzziness index. If $m = 1$, the algorithm is reduced to
the hard c-means algorithm. The necessary conditions for the minimizer
(U^*,V^*) of $J_m(U,V)$ are defined as

$$u_{ik} = \frac{1}{\sum_{j=1}^{c} \left(\frac{d(x_k, V_i)}{d(x_k, V_j)} \right)^{\frac{2}{(m-1)}}}, \tag{15.3}$$

$$V_i = \frac{\sum_{k=1}^{n} (u_{ik})^m x_k}{\sum_{k=1}^{n} (u_{ik})^m}, \tag{15.4}$$

where $d^2(x_k, V_i) = \|x_k - V_i\|^2$ is the distance from x_k to the cluster center V_i.

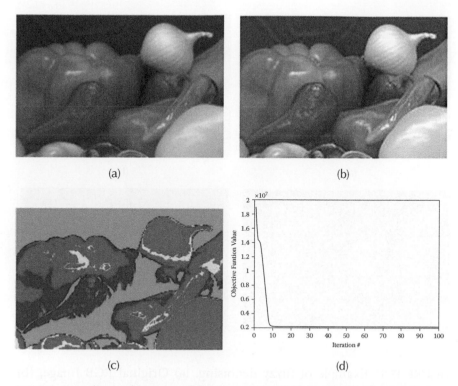

FIGURE 15.4 (See color insert following Page 204.) Example of using Fuzzy clustering technique for Image Classification. (a) Original RGB Image, (b) gray-scale version of (a), (c) classified image, and (d) objective function progression.

The classification algorithm used in Chapter 12 is modified in the following example for the fuzzy c-mean clustering technique. As such, the need for a training set is eliminated, and artificial measures based on the mean of a selected-class area need not be calculated. Figure 15.4 shows the application example.

15.2.3 Fuzzy Denoising

A gray-tone image taken of a real scene will contain inherent ambiguities due to light dispersion on the physical surfaces. The neighboring pixels may have very different intensity values and yet represent the same surface region. In this book, a fuzzy set theoretic approach to representing, processing, and quantitatively evaluating the ambiguity in gray-tone images is presented. The gray-tone digital image is mapped into a two-dimensional array of singletons called a fuzzy image. The value of each fuzzy singleton reflects the degree to which the intensity of the corresponding pixel

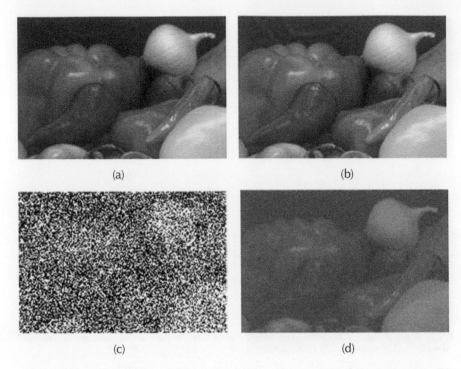

(a) (b)

(c) (d)

FIGURE 15.5 Example of fuzzy denoising. (a) Original RGB Image, (b) grayscale version of (a), (c) noisy image in 0 mean unit variance Gaussian noise, (d) filtered image.

is similar to the neighboring pixel intensities. The inherent ambiguity in the surface information can be modified by performing a variety of fuzzy mathematical operations on the singletons. Once the fuzzy image processing operations are complete, the modified fuzzy image can be converted back to a gray-tone image representation. The ambiguity associated with the processed fuzzy image is quantitatively evaluated by measuring the uncertainty present both before and after processing.

In previous chapters, certain types of image filters designed to remove noise were presented. Figure 15.5 shows the application of a fuzzy inference system (ANFIS) to remove noise from a supervised image. The implication is that a less noisy version of this image is used as a training set to fine-tune the membership functions to allocate new pixel values based on knowledge of the crude version of the image. This approach actually extracted a noise-free image from a highly noisy image.

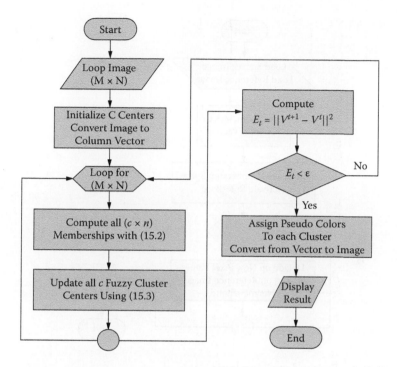

FIGURE 15.6 Algorithm for fuzzy clustering.

15.3 ALGORITHMIC ACCOUNT

The two examples presented in the previous section can be programmed according to the logic shown in Figures 15.6 and 15.7, respectively.

15.4 MATLAB CODE

In this section, the MATLAB code used to generate the images in Figures 15.4 and 15.5 is presented.

```
% Unsupervised Classification with Fuzzy C-Mean
% Only input from user is the number of classes (N)

close all
clear all

N = 5 ; % Number of classes
x=imread('onion.png');
figure ; imshow(x)
```

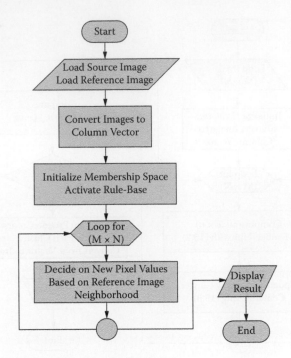

FIGURE 15.7 Fuzzy denoising logic.

```
y = rgb2gray(x);
figure ; imshow(y)
z = double(y) ;
[r,c] = size(y) ;

Z = im2col(z,[1 1])' ;
[center,U,obj_fcn] = fcm(Z,N);
c1 = sort(center)/max(center) ;
bn = (c1(2:N) - c1(1:N-1) ) / 2 + c1(1:N-1) ;
colors = [ 255 0 0 ; 0 255 0 ; 0 0 255 ; 255 255 0 ; 255
0 255 ; 0 255 255 ; 255 255 255 ; 0 0 0 ] ;
Y = x ;

for j = 1 : 3
    Y(:,:,j) = colors(N,j) ;
end

for i = 1 : length(bn)
    z = col2im(U(i,:),[1 1],[r c]) ;
    [A,B] = find(z>=bn(i)) ;
```

```
    for k = 1 : length(A)
        for j = 1 : 3
        Y(A(k),B(k),j) = colors(i,j) ;
   end
  end
end
figure ; imshow(Y)

% Filtering

close all
clear all
x=imread('onion.png');
figure ; imshow(x)
y = rgb2gray(x);
figure ; imshow(y)

z = imnoise(y,'gaussian',0,1) ; % actual noisy version
yn = imnoise(y,'gaussian') ; % low noise version
figure ; imshow(z)
y = double(y) ;
yn = double(yn) ;
z = double(z) ;
[r,c] = size(y) ;
Y = im2col(y,[1 1])' ;
YN = im2col(yn,[1 1])' ;
Z = im2col(z,[1 1])' ;

delayed_Y = [0; Y(1:length(Y)-1)];
trn_data = [delayed_Y YN Z];
% Generating the initial FIS
mf_n = 3;
ss = 0.3;
in_fismat=genfis1(trn_data, mf_n);
% Using ANFIS to finetune the initial FIS
out_fismat = anfis(trn_data, in_fismat, [nan nan ss]);
% Testing the tuned model with training data
estimated_n2 = evalfis(trn_data(:, 1:2), out_fismat);
estimated_x = Z - estimated_n2;

yf = col2im(estimated_n2,[1 1],[r c]) ;
figure ; imshow(yf*0.004)
```

15.5 SUMMARY

- Soft computing represents the nonconventional approach to solving engineering problems related to modeling, estimation, and pattern recognition.

- In general, the domain of soft computing encompasses fuzzy logic, neural networks, genetic and evolutionary algorithms, swarm intelligence, chaos theory, and similar techniques.

- Soft computing can be applied to image processing for specific tasks.

- Fuzzy logic has been presented in this chapter with applications to image classification and denoising.

- For classification, a fuzzy c-means clustering algorithm has been used.

- For denoising, a fuzzy inference system is used with a reference image for approximating the input image pixels.

- MATLAB's Fuzzy Logic Toolbox is used to implement the algorithms.

- Along the same lines, neural networks can also be used for the same two applications.

- Other methodologies of soft computing could also be used. However, which approach would be better would depend on the nature of the problem.

- Usually, the better approach is a hybrid technique, in order to make use of the best features of each method.

15.6 EXERCISES

1. Repeat the denoising application discussed in Section 15.2.3 with a Neural Network as main decision maker instead of neighborhood rules. Use the NEWPNN function for the main network. Figure 15.7 can be referred to for further clarification.

2. Use the following set of kernels and develop a set of if-then-else rules to decide whether a foreground pixel is 1.

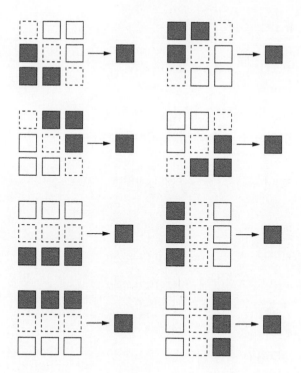

3. Implement Q #2 using the RBFNN function in MATLAB.

Bibliography

Abeyratne, U. R., Petropulu, A. P., and Reid, J. M. "Higher order spectra based deconvolution of ultrasonic image." *IEEE Transactions on Ultrasonics, Ferroelectrics, and Frequency Control* 42, no. 6 (1995): 1064–1075.

Ahn, H., and Yagle, A. E. "2D blind deconvolution by partitioning into coupled 1D problems using discrete Radon transforms." *Proceedings of the International Conference on Image Processing* (1995): 37–40.

Amari, S., Cichocki, A., and Yang, H. H. "A new learning algorithm for blind signal separation." In D. Touretzky, M. Mozer, and M. Hasseimo (Eds.) *Advances in Neural Information Processing Systems 8.* Cambridge, MA: MIT Press, 1996, pp. 757–763.

Arlt, B. and Brause, R. "The principal independent components of images." Internal Report 1/98, J. W.Goethe-Universität Frankfurt, Germany, 1998, http://www. informatik.uni-frankfurt.de/ fbreports/01_.98.ps.gz.

Ballard, D. H. "Generalizing the Hough transform to detect arbitrary shapes." *Pattern Recognition*, 13 (1981): 111–122.

Ballard, D. H., and Brown, C. *Computer Vision.* Englewood Cliffs, NJ: Prentice Hall, 1982.

Banham, M. R., and Katsaggelos, A. K. "Digital image restoration." *IEEE Signal Processing Magazine* (March 1997): 24–41.

Banham, M. R., and Katsaggelos, A. K. "Spatially adaptive wavelet-based multiscale image restoration." *IEEE Transactions on Image Processing* 5, no. 4 (1996): 619–634.

Barlow, H. B. "Unsupervised learning." *Neural Computation*, (1989): 295–311.

Bartlett, M. S. and Lades, H. M. "Sejnowski: Independent component representations for face recognition." *Proceedings of the SPI Symposium on Electronic Imaging* (in press) 1998.

Bennett, J., and Khotanzad, A. "Maximum likelihood estimation methods for multispectral random field image models." *IEEE Transactions on Pattern Analysis and Machine Intelligence*, 21, no. 6 (1999): 537–543.

Bernd J. *Practical Handbook on Image Processing for Scientific Applications.* CRC Press, New York, 1997.

Burns, J. B., Weiss, R. S., and Riseman, E. M. "View variation of point-set and line-segment features." *IEEE Transactions on Pattern Analysis and Machine Intelligence*, 15, no. 1(1993): 51–68.

Comon, P. "Independent component analysis—a new concept?" *Signal Processing*, 36 (1994): 287–314.

Cristianini, N., Shawe-Taylor, J., and Kandola, J. "Spectral kernel methods for clustering," in T. G. Dietterich, S. Becker, and Z. Ghahramani (Eds.) *Advances in Neural Information Processing Systems 14*. Cambridge MA: MIT Press, 2002.

Davies, E. R. *Machine Vision—Theory, Algorithms, Practicalities*. New York: Academic Press, 1990.

Demoment, G. "Image reconstruction and restoration: Overview of common estimation structures and problems." *IEEE Transactions on Acoustics, Speech, and Signal Processing*, 37, no. 12 (1989): 2024–2036.

Dhawan, A. P., Rangayyan, R. M., and Gordon, R. "Image restoration by Wiener deconvolution in limited-view computed tomography." *Applied Optics*, 24, no. 23 (1985): 4013–4020.

Drury, Stephen A. "Image Interpretation in Geology," Chap. 5 in *Digital Image Processing*, 3rd ed. New York: Routledge, 2001.

Duda, R. O. and Hart, P. E. "Use of the Hough transformation to detect lines and curves in pictures." *Communications of the ACM*, 15 (1972): 11–15.

Duda, R. O., Hart, P. E., and Stork, D. G. *Pattern Classification*, 2nd ed. New York: Wiley, 2001.

Fyfe, C., and Lai, P. L. "ICA using kernel canonical correlation analysis." In *International Workshop on Independent Component Analysis and Blind Signal Separation (ICA2000)*, Helsinki, June 2000.

Galatsanos, N. P., Katsaggelos, A. K., Chin, R. T., and Hillery, A. D. "Least squares restoration of multichannel images." *IEEE Transactions on Signal Processing* 39, no. 10 (1993): 2222–2236.

Gonzales, R. C., and Woods, R. E. *Digital Image Processing*, New York: Addison-Wesley Publishing Company, 1992.

Gonzales, R. C., Woods, R. E., and Eddins, S. L. *Digital Image Processing Using MATLAB®*, Upper Saddle River, NJ: Pearson Education Inc., 2004.

Harikumar, G., and Bresler, Y. "Exact image deconvolution from multiple FIR blurs." *IEEE Transactions on Image Processing* 8, no. 6 (1999): 846–862.

Harikumar, G., and Bresler, Y. "Perfect blind restoration of images blurred by multiple filters: Theory and efficient algorithms." *IEEE Transactions on Image Processing* 8, no. 2 (1999): 202–219.

Horn, R. A., and Johnson, C. R. *Matrix Analysis*. Cambridge, UK: Cambridge University Press, 1985.

Horn, R. A., and Johnson, C. R. *Topics in Matrix Analysis*. Cambridge, UK: Cambridge University Press, 1991.

Huang, T. S., Schreiber, W. F., and Tretiak, O. J. "Image processing." *Proceedings of the IEEE*, 59, no. 11(1971):1586–1609.

Javidi, B., Caulfield, H. J., and Horner, J. L. "Image deconvolution by nonlinear signal processing." *Applied Optics* 28, no. 15 (1989): 3106–3111.

Jensen, J. R. "Introductory Digital Image Processing: A Remote Sensing Perspective." Chap. 8 in *Thematic Information Extraction: Image Classification*. Upper Saddle River, NJ: Prentice Hall, 1996.

Kang, M. G. "Generalized multichannel image deconvolution approach and its applications." *Optical Engineering*, 37, no. 11 (1998): 2953–2964.

Kundur, D., and Hatzinakos, D. "Blind image deconvolution." *IEEE Signal Processing Magazine* May 1996: 43–64.

Lagendijk, R. L., Tekalp, A. M., and Biemond, J. "Maximum likelihood image and blur identification: a unifying approach." *Optical Engineering*, 29, no. 5 (1990): 422–435.

Lee, K. S., Kim, E. S., Doh, W., and Youn, H. "Image enhancement based on signal subspace approach." *IEEE Transactions on Image Processing* 8, no. 8 (1999): 1129–1134.

Lillesand, T. M. "Remote Sensing and Image Interpretation." Chap. 7 in *Digital Image Processing*, 5th ed. New York: John Wiley & Sons, 2004.

Lim, J. S. *Two-Dimensional Signal and Image Processing*. Upper Saddle River, NJ: Prentice Hall Signal Processing Series, 1990.

Maeda, J. "Image restoration by an iterative damped least-squares method with non-negativity constraint." *Applied Optics*, 24, no. 6 (1985): 751–757.

Malfait, M. and Roose, D. "Wavelet-based image denoising using a Markov random field a priori model." *IEEE Transactions on Image Processing* 6 no. 4 (1997): 549–565.

MATLAB Image Processing Toolbox. Reference Manual.

Mikhael, W. B., and Yu, H. "A linear approach for two-dimensional, frequency domain, least square, signal and system modeling." *IEEE Transactions on Circuits and Systems-II* 41, no. 12 (1994): 786–795.

Olshausen, B. A., and Field, D. J. "Natural image statistics and efficient coding." *Network: Computation in Neural Systems*, 7 (1996): 333–339.

Rosipal, R., and Trejo, L. J. "Kernel partial least squares regression in reproducing kernel Hilbert space." *Journal of Machine Learning Research*, 2 (2001): 97, 123.

Sandor, V. and Park, S. K. "Wavelet-based restoration for scenes with smooth bases." *Proceedings of the SPIE Conference on Applications of Digital Image Processing* 3460 (1998): 555–565.

Schalkoff, R., J., *Digital Image Processing and Computer Vision*. New York: John Wiley & Sons, 1989.

Schowengerdt, R. A. "Remote Sensing: Models and Methods for Image Processing," Chap. 9 in *Thematic Classification*. New York: Academic Press, 1997.

Scott, G., and Longuet-Higgins, H. "An algorithm for associating the features of two patterns." In *Proceedings of the Royal Society London B*, 224 (1991): 21, 26.

Sekko, E., Thomas, G., and Boukrouche, A. "A deconvolution technique using optimal Wiener filtering and regularization." *Signal Processing*, 72 (1999): 23–32.

Sezan, M. I., and Tekalp, A. M. "Image restoration and reconstruction." *Optical Engineering*, 29, no. 5 (1990): 391–392.

Shannon, C. E., and Weaver, W. *The Mathematical Theory of Information*. Urbana: University of Illinois Press, 1949.

Shawe-Taylor, J., and Cristianini, N. *Kernel Methods for Pattern Analysis.* Cambridge, U.K.: Cambridge University Press, 2004.

Shi, J., and Malik, J. "Normalized cuts and image segmentation." *IEEE Transactions on Pattern Analysis and Machine Intelligence,* 22, no. 8 (2000): 888, 905.

Taxt, T. "Representation of medical ultrasonic images using two-dimensional homomorphic deconvolution." *IEEE Transactions on Ultrasonics, Ferroelectrics, and Frequency Control* 42, no. 4 (1995): 543–554.

Tong, L., and Liu, R. "A closed form identification of multichannel moving average processes by ESPRIT." *Circuits and Systems Signal Processing* 15, no. 3 (1996): 343–359.

Vapnik, V. N. *The Nature of Statistical Learning Theory,* 2nd ed. New York: Springer, 1999.

Vogel, C. R., and Oman, M. E. "Fast, robust total variation-based reconstruction of noisy, blurred images." *IEEE Transactions on Image Processing* 7, no. 6 (1998): 813–824.

Weiss, Y. "Segmentation using eigenvectors: A unifying view." In *Proceedings of the 7th International Conference on Computer Vision* (Kerkyra, September 1999): 975–982.

Witten, H., Neal, M., and Cleary, G. "Arithmetic coding for data compression." *Communications of the ACM* 30, no. 6 (June 1987): 520–540.

Xu, Y., and Crebbin, G. "Image blur identification by using HOS techniques." *IEEE International Conference on Image Processing* 3 (1996): 77–80.

Yu, X., Hsu, C. S., Bamberger, R. H., and Reeves, S. J. "$H\infty$ Deconvolution filter design and its application in image restoration." *IEEE International Conference on Acoustic Speech and Signal Processing* (1995): 2611–2614.

Zadeh L. "Fuzzy sets", *Information and control.* 8, (1965): 338–353.

Zames, G., "Feedback and optimal sensitivity: Model reference transformation, multiplicative seminorms, and approximate inverse." *IEEE Transactions on Automatic Control* 23, (1981): 301–320.

Zha, H., Ding, C., G. M., He, X., and Simon, H. "Spectral relaxation for k-means clustering." In T. G. Dietterich, S. Becker, and Z. Ghahramani (Eds.) *Advances in Neural Information Processing Systems 14.* Cambridge, MA: MIT Press, 2002.

Zhou, Y., Chellappa, R., Vaid, A., and Jenkins, B. K. "Image restoration using a neural network." *IEEE Transactions on Acoustics, Speech, and Signal Processing,* 36, no. 7 (1988): 1141–1151.

RELATED PUBLICATIONS FROM THE AUTHORS

"2D-H_∞-based deconvolution for image enhancement with applications to ultrasonic NDE." *IEEE Signal Processing Letters* 9, no. 5 (May 2002): 157–159.

"An efficient hole-filling algorithm for C-scan enhancement." *AIP Conference Proceedings,* 615, no. 1 (May 25, 2002): 662–669.

"Binary Image coding using 1D chaotic maps." *Proceedings IEEE Region 5 Technical Conference,* New Orleans, April 2003, pp. 39–42.

"Binary image transformation using two-dimensional chaotic maps." *IEEE International Conference on Pattern Recognition* (August 2004): 823–826.

"Blind enhancement for ultrasonic C-scans using recursive 2-D H_∞-based state estimation and filtering." *"INSIGHT," Journal of British Institute of Nondestructive Testing* 44, no. 11 (November 2000): 737–741.

"Blind image restoration for ultrasonic C-scan using constrained 2D-HOS." *Proceedings IEEE ICASSP* no. 6, (Utah, May 2001): 3405–3408.

"Blind image-deconvolution for ultrasonic C-scans." *RQNDE* (Iowa, July 2000): 273–274.

"Blind-H_∞ deconvolution for ultrasonic C-scans: 1-D approach." *Journal of NDE* 21, no. 2 (June 2001): 40–46.

"Chaotic gray-level image transformation," *Journal of Electronic Imaging*, 14, no. 4 (2005): 043001.

"C-scan enhancement using recursive subspace deconvolution." *AIP Conference Proceedings*, 615, no. 1 (May 25, 2002): 655–661.

"Hybrid 2D H_∞-based blind enhancement for ultrasonic C-scans." *Proceedings of the 39th Annual Conference of British Institute of Non Destructive Testing*, (Buxton, England, September 2000): 137–142.

"Image encoding using chaotic maps and strange attractors." *Proceedings of SPIE/IS&T; Image Processing: Algorithms and Systems II*, 5014, Santa Clara, California, January 2003: 1–8.

"Infrared Image Enhancement using H_∞ bounds for surveillance applications." *IEEE Transactions on Image Processing* 17, no. 8 (August 2008): 1274–1282.

"Recent trends in 2D blind deconvolution for nondestructive evaluation." *Tamkang Journal of Science and Engineering* 5, no. 1 (March 2002): 49–58.

"Recent trends in 2D blind deconvolution for nondestructive evaluation." *IEEE International Conference on Pattern Recognition* (August 2002): 989–992.

WEB RESOURCES

"A Closer Look at Huffman Encoding," at http://www.rasip.fer.hr/research/compress/algorithms/fund/huffman/.

A collection of technical terminologies at http://bmrc.berkeley.edu/frame/research/mpeg/faq/mpeggloss.html.

A definition of Soft-Computing adapted from L.A. Zadeh http://www.soft-computing.de/def.html.

A good introduction to Arithmetic coding can be found in "Arithmetic Coding+Statistical Modelling=Data Compression," at http://dogma.net/markn/articles/arith/part1.htm.

Compression newsgroup at (news:comp.compression).

Fractal Image Encoding at http://inls.ucsd.edu/y/Fractals/. Contains a real LOT of interesting papers and software related to Fractal image encoding.

Image Compression at http://www.iee.et.tu-dresden.de/~franz/image1.html.

Image Labs International http://www.imagelabs.com/.

Wavelet Image Compression Kit at http://www.geoffdavis.net/dartmouth/wavelet/wavelet.html.

Glossary

Binary images: Binary images are images whose pixels have only two possible intensity values. They are normally displayed as black and white. Numerically, the two values are often 0 for black, and either 1 or 255 for white. Binary images are often produced by thresholding a grayscale or color image in order to separate an object in the image from the background. The color of the object (usually white) is referred to as the *foreground color*. The rest (usually black) is referred to as the *background color*. However, depending on the image that is to be thresholded, this *polarity* might be inverted, in which case the object is displayed with 0 and the background with a nonzero value.

Bit: An acronym for a binary digit. It is the smallest unit of information that can be represented. A bit may be in one of two states, on or off, represented by a zero or a one.

Bit map: A representation of graphics or characters by individual pixels arranged in rows and columns. Black and white require 1 bit, while high-definition color requires up to 32 bits.

Brightness: Magnitude of the response produced in the eye by light.

Byte: A group of eight bits of digital data.

Color images: It is possible to construct (almost) all visible colors by combining the three primary colors, red, green, and blue (RGB), because the human eye has only three different color receptors, each of which is sensitive to one of the three colors. Full RGB color requires that the intensities of the three color components be specified for each and every pixel. It is common for each component intensity to be stored as an 8-bit integer, and so each pixel requires 24 bits to completely and accurately specify its color. If this is done, the image is known as a 24-bit color image.

Color depth: The number of color values that can be assigned to a single pixel in an image. Also known as bit depth, color depth can range from 1 bit (black and white) to 32 bits (over 16.7 million colors).

Color quantization: *Color quantization* is applied when the color information of an image is to be reduced. The most common case is when a 24-bit color image is transformed into an 8-bit color image.

Contrast: The difference between highlights and shadows in a photographic image. The larger the difference in density, the greater the contrast.

Contrast stretching: Improving the contrast of images by digital processing. The original range of digital values is expanded to utilize the full contrast range of the recording film or display device.

Convolution: Convolution is a simple mathematical operation that is fundamental to many common image processing operators. Convolution provides a way of "multiplying together" two arrays of numbers, generally of different sizes, but of the same dimensionality, to produce a third array of numbers of the same dimensionality. This can be used in image processing to implement operators whose output pixel values are simple linear combinations of certain input pixel values. The convolution is performed by sliding the kernel over the image, generally starting at the top-left corner, so as to move the kernel through all the positions where the kernel fits entirely within the boundaries of the image. (Note that implementations differ in what they do at the edges of images, as explained later.) Each kernel position corresponds to a single output pixel, the value of which is calculated by multiplying together the kernel value and the underlying image pixel value for each of the cells in the kernel, and then adding all these numbers together.

Correlation: A mathematical measure of the similarity between images or areas within an image. Pattern matching or correlation of an X-by-Y array size template to an image of the same size produces a scalar number, the percentage of match. Typically, the template is walked through a larger array to find the highest match.

Digital watermark: A unique identifier embedded in a file to deter piracy and prove file ownership and quality.

Dilation: A morphological operation that moves a probe or structuring element of a particular shape over the image, pixel by pixel. When an object boundary is contacted by the probe, a pixel is preserved in the output image. The effect is to "grow" the objects.

Distance metrics: It is often useful in image processing to be able to calculate the distance between two pixels in an image, but this is not as straightforward as it seems. The presence of the pixel grid makes several so-called *distance metrics* possible, which often give different answers for the distance between the same pair of points. The three most important ones are the following:

1. Euclidean distance—The straight line distance between two pixels at coordinates (x_1, y_1) and (x_2, y_2); the Euclidean distance is given by $\sqrt{(x_2 - x_1)^2 + (y_2 - y_1)^2}$.

2. City block distance—Also known as the Manhattan distance. This metric assumes that in going from one pixel to the other, it is only possible to travel directly along pixel grid lines. Diagonal moves are not allowed. Therefore, the city block distance is given by $|x_2 - x_1| + |y_2 - y_1|$.

3. Chessboard distance—This metric assumes that you can make moves on the pixel grid as well as diagonally with equivalent distance. This means that the metric is given by $(|x_2 - x_1| + |y_2 - y_1|)$.

Dithering: Dithering is an image display technique that is useful for overcoming limited display resources. The word *dither* refers to a random or semirandom perturbation of the pixel values.

Edge: A change in pixel values exceeding some threshold amount. Edges represent borders between regions on an object or in a scene.

Edge detection: The ability to determine the edge of an object.

Edge enhancement: The process of identifying edges or high frequencies within digital images.

Erosion: The converse of the morphology dilation operator. A morphological operation that moves a probe or structuring element of a particular shape over the image, pixel by pixel. When the probe fits inside an object boundary, a pixel is preserved in the output image. The effect is to "shrink" or "erode" objects as they appear in the output image. Any shape smaller than the probe (i.e., noise) disappears.

Fast Fourier transform: Produces a new image that represents the frequency domain content of the spatial or time domain image information. Data is represented as a series of sinusoidal waves.

Frame grabber: This device is used to convert analog signals to digital signals; used in digital imaging.

Grayscale images: A grayscale (or gray-level) image is one in which the only colors are shades of gray. The reason for differentiating such images from any other type of color image is that less information needs to be provided for each pixel. In fact, a "gray" color is one in which the red, green, and blue components all have equal intensity in RGB space, and so it is only necessary to specify a single intensity value for each pixel, as opposed to the three intensities needed to specify each pixel in a full-color image. Often, the grayscale intensity is stored as an 8-bit integer, giving 256 possible different shades of gray, from black to white. If the levels are evenly spaced, then the difference between successive gray levels is significantly better than the gray-level resolving power of the human eye.

High-pass filter: Allows detailed high-frequency image information to pass while attenuating low-frequency, slow-changing data. Opposite of low-pass filter.

Histogram: A graphical representation of the frequency of occurrence of each intensity or range of intensities (gray levels) of pixels in an image. The height represents the number of observations occurring in each interval.

Histogram equalization: Modification of the histogram to evenly distribute a narrow range of image grayscale values across the entire available range.

Hue: The attribute of a color that differentiates it from gray of the same brilliance and that allows it to be classed as blue, green, red, or intermediate shades of these colors.

Image: Projection of an object or scene onto a plane (i.e., screen or image sensor).

Image analysis: The process of extracting features or attributes from an image based on properties of the image; evaluation of an image based on its features, for decision making.

Image capture/acquisition: The process of acquiring an image of a part or scene from sensor irradiation to acquisition of a digital image.

Image distortion: A situation in which the image is not exactly true to scale with the object scale.

Image enhancement: Image processing operations that improve the visibility of image detail and features. Usually performed either automatically by software or manually by a user through an interactive application. Any one of a group of operations that improves the detectability of the targets or categories. These operations include, but are not limited to, contrast improvement, edge enhancement, spatial filtering, noise suppression, image smoothing, and image sharpening.

Image filter: A mathematical operation performed on a digital image at every pixel value to transform the image in some desired way.

Image processing: Conversion of an image into another image in order to highlight or identify certain properties of the image.

Image resampling: A technique for geometric correction in digital image processing. Through a process of interpolation, the output pixel values are derived as functions of the input pixel values combined with the computed distortion. Nearest neighbor, bilinear interpolation, and cubic convolution are commonly used resampling techniques.

Kernel: A kernel is (usually) a small matrix of numbers that is used in image convolutions. Different-sized kernels containing different patterns of numbers give rise to different results under convolution. The word *kernel* is also commonly used as a synonym for *structuring element*, which is a similar object used in mathematical morphology.

Masking: A mask is a binary image consisting of both zero and nonzero values. If a mask is applied to another binary or to a grayscale image of the same size, all pixels that are zero in the mask are set to zero in the output image. All other pixels remain unchanged. Masking can be implemented using either pixel multiplication or logical AND, the latter in general being faster. Masking is often used to restrict a point or arithmetic operator to an area defined by the mask. We can, for example, accomplish this by first masking the desired area in the input image and processing it with the operator, then masking the original input image with the inverted mask to obtain the unprocessed area of the image and, finally, recombining the two partial images using image addition. In some image processing packages, a mask can directly be defined as an optional input to a point operator, so that the operator is automatically applied only to the pixels defined by the mask.

Mean squared error: The mean squared error is a measure of performance of a point estimator. It measures the average squared difference between the estimator and the parameter. For an unbiased estimator, the mean squared error is equal to the variance of the estimator.

Median: In a population or a sample, the median is the value that has just as many values above it as below it. If there are an even number of values, the median is the average of the two middle values. The median is a measure of central tendency. The median can also be defined as the 50th percentile. For symmetrical distributions, the median coincides with the mean and the center of the distribution. For this reason, the median of a sample is often used as an estimator of the center of the distribution. If the distribution has heavier tails than the normal distribution, then the sample median is usually a more precise estimator of the distribution center than is the sample mean.

Median filter: A method of image smoothing that replaces each pixel value with the median grayscale value of its immediate neighbors.

Mode: The mode is a value that occurs with the greatest frequency in a population or a sample. It could be considered as the single value most typical of all the values.

Morphology: Group of mathematical operations based on manipulation and recognition of shapes. The study of shapes and the methods used to transform or describe shapes of objects. Also called mathematical morphology. Operations may be performed on either binary or grayscale images.

Noise: Random or repetitive events that obscure or interfere with the desired information.

Noisy image: An image with many pixels of different intensities. An untuned TV picture produces a very noisy or random image. (Note that sound has nothing to do with a noisy image.)

Normal distribution: The normal distribution is a probability density that is bell-shaped, symmetrical, and single-peaked. The mean, median, and mode coincide and lie at the center of the distribution. The two tails extend indefinitely and never touch the x-axis (asymptotic to the x-axis). A normal distribution is fully specified by two parameters: the mean and the standard deviation.

Pattern recognition: A process of decision making in which a new input is recognized as a member of a given class by a comparison of its attributes with the already known pattern of common attributes or members of that class.

Picture element (pixel): In a digitized image, this is the area on the ground represented by each digital value. Because the analog signal from the detector of a scanner may be sampled at any desired interval, the picture element may be smaller that the ground resolution cell of the detector. Commonly abbreviated as pixel.

Principal components analysis: The purpose of principal components analysis is to derive a small number of linear combinations (principal components) of a set of variables that retain as much of the information in the original variables as possible. This technique is often used when there are large numbers of variables, and you wish to reduce them to a smaller number of variable combinations by combining similar variables (ones that contain much the same information). Principal components are linear combinations of variables that retain maximal amount of information about the variables. The term "maximal amount of information" here means the best least-squares fit, or, in other words, maximal ability to explain variance of the original data.

Resolution: The ability to distinguish closely spaced objects on an image or photograph. Commonly expressed as the spacing, in line pairs per unit distance, of the most closely spaced lines that can be distinguished.

RGB: Acronym for red–green–blue. A model for describing colors that are produced by emitting light, as on a video monitor, rather than by absorbing it, as with ink on paper. The three kinds of cone cells in the eye respond to red, green, and blue light, respectively, so percentages of these additive primary colors can be mixed to get the appearance of any desired color.

Segmentation: The process of dividing a scene into a number of individual objects or contiguous regions, differentiating them from each other and the image background.

Template: An artificial model of an object or a region or feature within an object.

Template matching: A form of correlation used to find out how well two images match.

Texture: The degree of smoothness of an object surface. Texture affects light reflection, and is made more visible by shadows formed by its vertical structures.

Thresholding: The process of converting a grayscale image into a binary image. If the pixel's value is above the threshold, it is converted to white. If it is below the threshold, the pixel value is converted to black.

Index

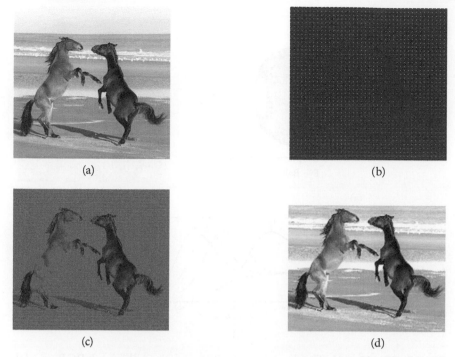

COLOR FIGURE 1.1   Digitization of a continuous image. (a) Original image of size $391 \times 400 \times 3$, (b) image information is almost lost if sampled at a distance of 10 pixels, (c) resultant pixels when sampling distance is 2, (d) new image with 2-pixel sampling with size $196 \times 200 \times 3$.

COLOR FIGURE 1.3   The effect of bit-resolution. (a) Original image with 8-bit resolution, (b) 4-bit resolution, (c) 3-bit resolution, (d) 2-bit resolution.

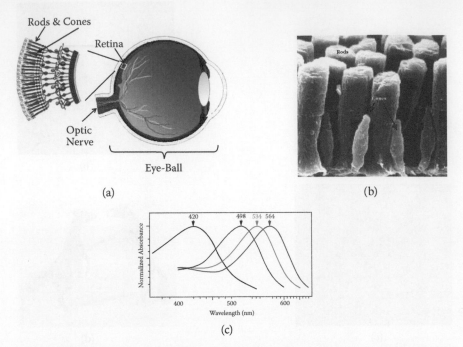

COLOR FIGURE 2.4 The human eye, its sensory cells, and RGB model sensitivity curves. (a) Diagram of the human eye showing various parts of interest, (b) cells of rods and cones under the electron microscope, (c) experimental curves for the excitation of cones for different frequencies.

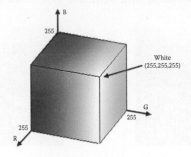

COLOR FIGURE 2.5 RGB color cube for the 8-bit unsigned integer representation system.

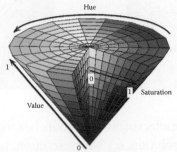

COLOR FIGURE 2.6 Illustration of the HSV color space.

0.0754	0.0588	0.0078
0.5137	0.5020	0.1843
0.2745	0.2980	0.1490
0.8667	0.7882	0.4353
0.4275	0.4824	0.6039
0.1882	0.2980	0.3686
0.1882	0.2078	0.0941
0.3214	0.3255	0.1490
0.3686	0.2235	0.0549
0.4118	0.3412	0.8157
0.3294	0.3843	0.4078

45	105	50	3	62	86	3	86	12	62
79	50	62	62	3	86	86	86	86	12
94	89	68	3	3	62	86	86	86	12
79	51	44	99	62	3	86	86	86	62
45	127	89	28	1	99	62	86	86	86
25	89	28	33	33	50	1	86	86	62
8	50	99	50	33	33	7	99	99	3
105	105	105	28	8	50	83	105	119	121
62	12	105	8	83	68	22	60	68	103
105	86	58	12	12	92	12	16	16	42

(a)

COLOR FIGURE 2.8   Anatomy of image types. (a) Index image with 128 color levels showing the values of indices and the map for a selected area in the image, (b) index image with 8 color levels, (c) grayscale image for index image with 8 levels.

(a)   (b)

(c)   (d)

COLOR FIGURE 2.11   Original RGB image with its components. (a) Source RGB image, (b) red component, (c) green component, (d) blue component.

(a)

(b)

(c)

(d)

COLOR FIGURE 4.1 Various affine operations. (a) Original image, (b) translated by (±240, ±320), (c) rotated by 45°, (d) scaled by 25%.

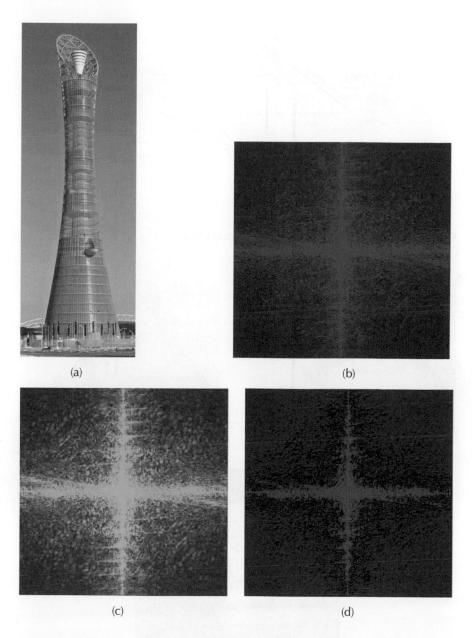

(a)

(b)

(c)

(d)

COLOR FIGURE 5.1   Application of Fourier transform to images. (a) Original
RGB image, (b) R, (c) G, (d) B components of the Fourier-transformed
image.

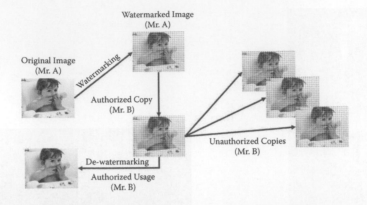

COLOR FIGURE 11.1   Concept of watermarking.

(a)

(b)

(c)

COLOR FIGURE 12.1   Image classification using the grayscale ranges only.
(a) Original RGB image, (b) grayscale version of (a), (c) segmented image.

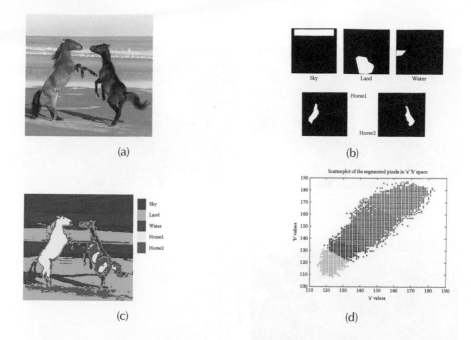

(a)

(b)

(c)

(d)

COLOR FIGURE 12.3 Application of nearest neighbor algorithm to classify the given image into its classes. (a) Sample image, (b) selected areas as training classes, (c) segmented image, (d) class boundaries.

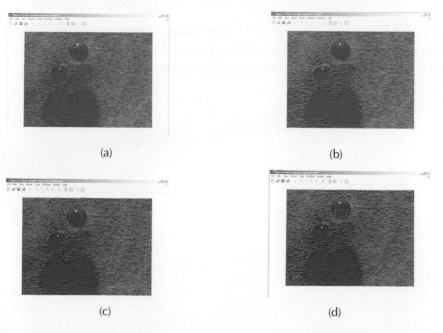

(a)

(b)

(c)

(d)

COLOR FIGURE 13.6 Declaring various classes for color-based tracking. (a) Target 1 is highlighted for red color, (b) target 2 is highlighted for blue color, (c) background is highlighted, (d) dark areas are highlighted.

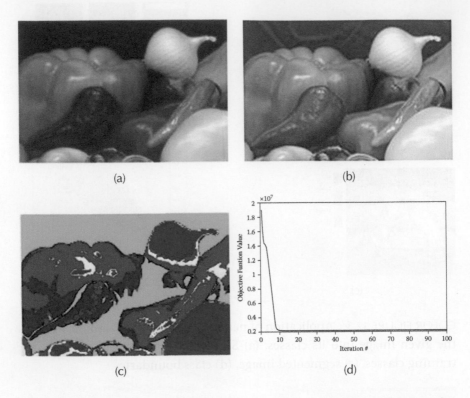

(a)

(b)

(c)

(d)

COLOR FIGURE 15.4 Example of using Fuzzy clustering technique for image classification. (a) Original RGB Image, (b) grayscale version of (a), (c) classified image, and (d) objective function progression.

Printed and bound by CPI Group (UK) Ltd, Croydon, CR0 4YY

18/10/2024

01776267-0004